Botanical Drug Products

Botanical Drug Products
Recent Developments and Market Trends

Edited by

Jayant N. Lokhande

Yashwant Pathak

CRC Press
Taylor & Francis Group
Boca Raton London New York

CRC Press is an imprint of the
Taylor & Francis Group, an **informa** business

CRC Press
Taylor & Francis Group
6000 Broken Sound Parkway NW, Suite 300
Boca Raton, FL 33487-2742

First issued in paperback 2020

© 2019 by Taylor & Francis Group, LLC
CRC Press is an imprint of Taylor & Francis Group, an Informa business

No claim to original U.S. Government works

ISBN-13: 978-1-4987-4005-0 (hbk)
ISBN-13: 978-0-367-73247-9 (pbk)

Visit the Taylor & Francis Web site at
http://www.taylorandfrancis.com

and the CRC Press Web site at
http://www.crcpress.com

This book is dedicated to our Family Members, Sages & Mystical Masters, Friends, Fellow Colleagues, Mother Nature, and Scientists who have always inspired us to innovate technologies and applications for quality-of-life improvement.

Contents

Foreword

Kalyuga, the era of machines and technologies of twenty-first century, has brought a revolution in the life of human beings. Such technologies are constantly improving the health and longevity of human life and have helped explore new frontiers in the field of biomedical research. On the other hand, new health problems including human developmental disorders, acute and chronic diseases, quality of life, environmental changes, and socio-politico-economical-cultural instability have also been recognized as challenges for the human race.

In healthcare there is paucity of new and effective drug discovery for complex health disorders. Pharmaceutical industries and Academic Institutions are actively exploring new drug discovery and business models to maximize returns on resources consumed. The need of today is to reevaluate ancient wisdom, human experiences embedded in history with the help of modern technologies, and derive new thinking models so that sustainable technologies and applications can be designed.

In such scenario, ancient healthcare experiences and medical literature is promising as it offers new drug development strategies when coupled with new analytical techniques. Almost more than 90% of current lifesaving and common drugs synthesized in Pharmaceutical companies have origin somewhere from plants or other live organisms. Hence, instead of chemical synthesis of these drugs, if they are prepared directly from natural sources, the drug development cost can be reduced significantly. Also, other ingredients present in the natural source may be beneficial in increasing the efficacy of the active ingredient and/or also may reduce the side effects. single drug/single receptor/single action pharmaceutics theory of today is failing in treatments of complex disorders like cancer, cardiovascular diseases and chronic metabolic diseases, just to mention a few. The remedy in this conundrum is Poly Drug-Poly Action therapeutics, which already exists in Botanical Drug Products.

"Botanical Drug Product—Recent Developments and Market Trends," consisting of 12 chapters, is a propitious book as it explains various facets of Botanical Drug Development Process, mandatory tools and techniques, allied regulatory approval aspects, applicable therapeutic areas and respective market and intellectual property acquisition to have market exclusively. The scientists and physicians working in industry and biomedical academic institutes can plan basic and translational research together in Botanical Drug Production with its realistic risk management as responsible efforts to proffer safe and effective drugs for formidable diseases. For that, they should welcome the plethora of information collected in this monumental book. I sincerely think this book may also be a valuable source of information for younger generation students who wish to venture and adopt their carrier in pharmaceutical research.

Sen Pathak
Department of Genetics
University of Texas

Preface

The key points to contemplate while deciding the future of modern drug discovery and development are "Do we have right now or in the near future new drugs for modern ailments?" and "Can better quality and longer length of life be expected out of these new drugs?" The answers to both points are uncertain, at least on the basis of past 20 years data of new drug approvals and iatrogenic diseases.

Also, drugs are failing in terms of providing satisfactory disease management that may be due to differing genotypes and phenotypes of disease population. The pathogenesis and etiology of more than 50% of diseases are still poorly understood, notwithstanding newer analytical science and techniques employed. That's why scientific instances are exhibiting thinking compulsion, "Are we unsuccessful in understanding health in its totality?"

In such framework of drug discovery and development landscape, we strongly sense that both Industry and Academia has to re-strategize thinking and explore integrated avenues in order to have new knowhow of healthcare system. Integrating Science of Nature, Human Civilizational Data, historical dynamics of healthcare management with modern analytical technologies can result in absolutely new scientific concepts and applications in drug discovery and development.

Modern drugs may be based on chemistry or biology, but they have primordial origins from nature integrals such as Plants, Fossils etc. So we have in general rediscovered what already exists in Universe.

Botanical Drug Products (BDP) can contribute in managing Complex Diseases in two ways—one by treating the disease from its original cause rather than just managing it, and second improving quality of life with better lifestyles thereafter. Botanical Drug Products differs from Botanical Originated Conventional Drugs. Botanical Drug Products are Poly-Molecules in nature whereas Botanical Originated Conventional Drugs are Single-Molecule in nature. Most of Complex disorders where in Conventional Drugs are failing mandate new approach of multi receptor targeting. Botanical Drug Products, by default, feature to target multi receptors and deal with disease condition in its totality. Once BDPs safety is proven, efficacy can be established on efficient clinical trial models like Evidence-Based Medicines through Translational Research. The advantage of such clinical trials models is one can actually measure new drugs impact on health outcome at large rather than just its concise clinical trial success. Globally Regulatory Agencies of respective countries are modifying frameworks for Botanical Drug Product approvals along with innovative clinical trial systems and methods of quality standardizations.

This Book in above context has addressed all critical aspects of Botanical Drug Development Process, its regulatory framework along with intellectual property management. In detail explanation of Preclinical development and where some study points can be waived by producing historical data on its safety is also highlighted. The consideration of Chemistry, Manufacturing, and Control procedures with various processes scale-up manufacturing techniques is also represented systematically. How integrated approaches like Reverse Pharmacology, Observational Therapeutics, and Knowledge derived from sustainably practiced medicinal systems like Ayurveda can bring new frontiers in disease treatments are vividly explained as well. A special focus on Computational Approaches in studying various research databases and predetermine Botanical Drug Products efficacy and safety is also elucidated. Successful Botanical Dug Products so far approved and in the pipeline are mentioned in this book as Case Studies.

We sincerely hope this book "Botanical Drug Products: Recent Developments and Market Trends" can trigger new thinking and discussion in among scientific and industrial community to explore newer types of drugs and promise quality health expectations at consumer end.

Editors

Jayant N. Lokhande, M.D. (Botanical Drugs): MBA—Biotechnology is an expert in Botanical Drugs and Biotechnology Business Management. He has successfully strategized and formulated products for several nutraceutical, pharmaceutical, and medical food companies in the United States and other developed countries. He has significant clinical experience, especially in using Botanical Drugs and Medical Foods for Complex and Chronic Diseases. His other professional interests are in Biodiversity Entrepreneurship, Bioprospecting, Medical Anthropology, and Disease Reversal Therapeutics.

He is currently working as Chief Scientific Officer and Management Analyst in Indus Extracts in the US, and his responsibilities include Business Analysis and Product and Technical Market Development.

Yashwant Pathak received a PhD in pharmaceutical technology from Nagpur University, India, and EMBA and MS in conflict management from Sullivan University, in Kentucky. He is a professor and associate dean for faculty affairs at the College of Pharmacy, University of South Florida, in Tampa. With extensive experience in academia and industry, he has more than one hundred publications; two patents and two patent applications; and sixteen edited publications, including seven in nanotechnology and six in nutraceuticals and drug delivery systems. Also, he has several books on the subjects of cultural studies and conflict management. He has received several national and international awards. He was a recent Fulbright Senior Fellow in Indonesia for six months. He has traveled in seven countries and has lectured at more than 20 universities. He was a keynote speaker at the World Hindu Wisdom meeting in Bali, Indonesia, in June 2017, and co-shared the meeting with the governor of Bali.

Contributors

Ashika Advankar
National Institute of
   Pharmaceutical Education
   and Research (NIPER)
   – Ahmedabad
Gujarat, India

Grace Checo
MPH in Infectious Disease
   Management
USF College of Pharmacy
Undergraduate Research
University of Pittsburgh
Pittsburgh, Pennsylvania

Kareem Elgendi
College of Pharmacy
University of South Florida
Tampa, Florida

Markie Esmailian
PhyloSense
Consultant, NaturAI, Inc
New Jersey, USA

Kavita Gupta
Deputy Manager, Micro Labs
Pune, India

Anas Hanini
College of Pharmacy
University of South Florida
Tampa, Florida

Vishal Katariya
Founder, Katariya & Associates
Pune, India

Slavko Komarnytsky
Plants for Human Health
   Institute
North Carolina State University
Kannapolis, North Carolina

and

Department of Food,
   Bioprocessing, and Nutrition
   Sciences
North Carolina State University
Raleigh, North Carolina

Suma Krishnaswamy
CEO, Cambium Biotechnology
Bangalore, India

Kaushik Kuche
National Institute of
    Pharmaceutical Education and
    Research (NIPER) – Ahmedabad
Gujarat, India

Jayant N. Lokhande
Chief Scientific Officer, Indus
    Extracts
Los Angeles, California

Sonali Lokhande
Consultant, Criterion Edge
San Luis Obispo, California

Sameer Mahajan
Physician, Internal Indian System
    of Medicine
Mumbai, India

Rahul Maheshwari
National Institute of
    Pharmaceutical Education and
    Research (NIPER) – Ahmedabad
Gujarat, India

Ava Milani
College of Pharmacy
University of South Florida
Tampa, Florida

Adam Mohamed
College of Pharmacy
University of South Florida
Tampa, Florida

Aniko Nagy
Envision Biotechnology
Data Curator
Telki, Hungary

Kimberly Palatini
Plants for Human Health
    Institute
North Carolina State University
Kannapolis, North Carolina

and

Department of Food,
    Bioprocessing, and Nutrition
    Sciences
North Carolina State University
Raleigh, North Carolina

Param Patel
College of Pharmacy
University of South Florida
Tampa, Florida

Yashwant Pathak
College of Pharmacy
University of South Florida
Tampa, Florida

Timea Polgar
Envision Biotechnology
Consultant, NaturAI, Inc
New Jersey, USA

Achal Shah
College of Pharmacy
University of South Florida
Tampa, Florida

Piyoosh Sharma
TIT College of Pharmacy
Madhya Pradesh, India

Ganesh Shinde
Physician, Internal Indian System
of Medicine
Pune, India

Namrata Soni
Sam Higginbottom University of
Agriculture, Technology and
Sciences
Uttar Pradesh, India

Muktika Tekade
Sri Aurobindo Institute of Pharmacy
Madhya Pradesh, India

Rakesh Kumar Tekade
National Institute of
Pharmaceutical Education and
Research (NIPER) – Ahmedabad
Gujarat, India

Global Pharmaceuticals Industry and Botanical Drug Products

Jayant N. Lokhande and Sonali Lokhande

Contents

1.1 Global pharmaceuticals market updates

1.1.1 Market size and driving factors

The global pharmaceutical market is burgeoning. From gross sales of $1.08 trillion in 2011, sales are projected to settle at $1.6 trillion in 2020 with yearly average growth of 8%. The developed markets like United States, European Union, Japan, and Canada have exhibited comparatively lagging growth rate than soaring economy markets like Brazil, China, India, and Russia (referred to as the BRIC countries) where the pharmaceutical market is making headway at the growth rate of 22.6% (Figure 1.1).

This perpetual market growth is due to an increased life expectancy along with rapidly changing patterns of population behavior. In 2010, estimated global population was 6.9 billion; by 2020, it is presumed to be 7.6 billion. And, if current lifestyle trends continue, a significant proportion of the population will have health problems [2].

More than 30% of the population won't get enough physical exercise [3]. Twenty percent of the population and more will be overweight or obese [4] whereas more than 13% of population will be 60 or older [2]. There are other lifestyle factors that increase the risk of developing heart disease, diabetes, and cancer. The number of people reaching real old age is also mounting, and the prevalence of dementia doubles every 5 years after the

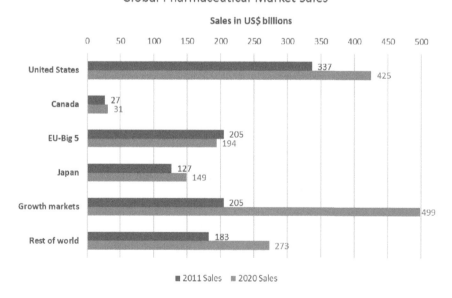

Figure 1.1 *Global pharmaceutical market size. (From Pharma 2020: From vision to decision, Available at https://www.pwc.com/gx/en/industries/pharmaceuticals-life-sciences/publications/pharma-2020.html [1].)*

age of 65 [5]. The World Health Organization (WHO) predicts that by 2020, non-communicable diseases will account for 44 million deaths a year—that's 15% more than in 2010 [3].

Globally infectious diseases are increasing as well, in part because some diseases have become drug-resistant. But over the past few decades new pathogens such as HIV and MRSA have emerged. And old scourges like pertussis have reared their heads again. In fact, the number of cases of pertussis in the U.S. is now higher than at any time since the early 1970s.

Quintiles have quoted that by 2021; global spending on branded pharmaceuticals and generics would culminate to $832 billion, with $495 billion U.S. dollars alone [6].

The research-based pharmaceutical industry is estimated to have spent nearly $149.8 billion U.S. dollars globally on pharmaceutical research and development (R&D) in 2015 [7].

In the United States, R&D investments of pharmaceutical companies have grown consistently over the past 15 years. From 2014 the investment has almost twice increased in the publicly funded organizations such as National Institutes of Health's (NIH) expenditures. Part of the U.S. Department of Health and Human Services, the NIH is the U.S. medical research agency, funding universities and research institutions in the United States and around the globe [8].

In 2015, the pharmaceutical industry registered 7,691 patents through the Patent Cooperation Treaty (PCT) of the World Intellectual Property Organization [9].

1.1.2 Diseases of future concern

1.1.2.1 Diseases of concern to human kind and industry

There is a huge disparity in between disease mortality causes in population and new drugs getting developed by industry as shown in Tables 1.1 and 1.2. According to the World Health Organization, circulatory system diseases represented the top two causes of mortality in the world in 2004, accounting for 23.6% of all deaths. Additionally, diseases of the respiratory system were the third and fourth leading causes of death worldwide in 2004 (a combined 12.1%). Given worldwide mortality rates, one would expect that the most-needed drugs would be those developed to treat diseases of the circulatory and respiratory systems; however, empirically it is a fact that a surprisingly small number of drugs are under development for circulatory system disorders ($n = 147$, 5.93%) and diseases of the respiratory system ($n = 168$, 6.78%) as shown in Figure 1.2. In fact, new drug development is more in the area of oncology, as shown in Table 1.2 and Figure 1.3. While specific cancers ranked as the 8th, 17th, and 20th disease-based causes of mortality in 2004, these diseases nonetheless represented fewer than 5% of deaths worldwide [10].

Table 1.1 Top 20 Global Disease Burdens

No.	Leading Causes of Death in 1990	Leading Causes of Death in 2006	Leading Causes of Death in 2016
1.	Ischemic heart disease	Ischemic heart disease	Ischemic heart disease
2.	Cerebrovascular disease	Lung cancer	Lung cancer
3.	Lung cancer	Cerebrovascular disease	Cerebrovascular disease
4.	Road injuries	Self-harm	Alzheimer's disease
5.	Self-harm	Alzheimer's disease	Self-harm
6.	Colorectal cancer	Road injuries	COPD
7.	COPD	Colorectal cancer	Colorectal cancer
8.	Alzheimer's disease	COPD	Lower respiratory infections
9.	Lower respiratory infections	Lower respiratory infections	Road injuries
10.	Stomach cancer	Breast cancer	Breast cancer
11.	Breast cancer	Diabetes	Diabetes
12.	Congenital anomalies	Stomach cancer	Pancreatic cancer
13.	Diabetes	Pancreatic cancer	Chronic kidney disease
14.	Neonatal preterm birth	Chronic kidney disease	Stomach cancer
15.	Other cardiovascular diseases	Liver cancer	Liver cancer
16.	Pancreatic cancer	Other cardiovascular diseases	Other cardiovascular disease
17.	Chronic kidney disease	Congenital anomalies	Drug use disorders
18.	Interpersonal violence	Leukemia	Other neoplasms
19.	Leukemia	Drug use disorders	Leukemia
20.	HIV/AIDS	Other neoplasms	Prostate cancer

Source: Naghavi, M. et al., *Lancet*, 390, 1151–1210, 2017 [10].

Table 1.2 New Drugs Being Developed for Various Disease Categories

Disease Category	New Drugs in Development
Oncology	1919
Neurology	1308
Infectious diseases	1261
Immunology	1123
Mental disorders	510
Cardiovascular diseases	563
Diabetes	401
HIV	208

Source: IFPMA, *IFPMA Facts and Figures Report: IFPMA*, IFPMA, 2017 [11].

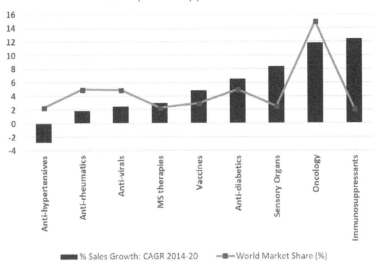

Figure 1.2 *Estimated market share and sales growth of new drug therapies in 2014–2020. (Adapted from Evaluate Pharma, World Preview 2015, Outlook to 2020, Evaluate Pharma, 2015 [12].)*

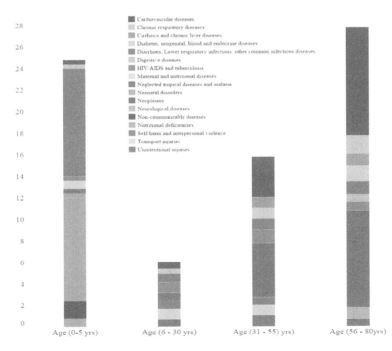

Figure 1.3 *Global composition of number of deaths for 4 age groups, both sexes combined, 1990 versus 2016. (From Naghavi, M. et al., Lancet, 390, 1151–1210, 2017 [10].)*

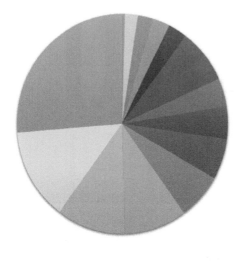

Percentage of unique drugs (n = 2477) developed in each therapeutic area

■ Complications of pregnancy, childbirth and the puerperium (0%)
■ Certain conditions originating in the perinatal period (0%)

■ Congenital anomalies (0.2%)

■ Symptoms, signs and ill-defined conditions (2 %)

■ Diseases of the blood and blood-forming organs (2%)

■ Injury and poisoning (3%)

■ Diseases of the genitourinary system (3%)

■ Diseases of the skin and subcutaneous tissue (4%)

■ Diseases of the digestive system (4%)

■ Mental disorders (4%)

■ Diseases of the musculoskeletal system and connective tissue (5%)
■ Diseases of the circulatory system (6%)

■ Diseases of the respiratory system (7%)

■ Endocrine, nutritional and metabolic diseases and immunity disorders (10%)
■ Infectious and parasitic diseases (11%)

■ Diseases of the nervous systems and sense organs (14%)

■ Neoplasm (26%)

Figure 1.4 *The percentage of drugs developed in various therapeutic areas. (From Cottingham, M.D. et al., Clin. Transl. Sci., 7, 297–299, 2014 [13].)*

As shown in Figure 1.4, the sample size included 2,477 unique drug entities in 4,182 clinical trials. The majority of drugs targeted neoplasms (26.20%), neurological diseases/diseases of the sense organs (13.48%), infectious and parasitic diseases (10.5%), and endocrine, metabolic, nutrition, and immunity disorders (9.45%). Less than 6% of drugs targeted diseases of the circulatory system, which represent the most prevalent causes of global mortality. Detailing the pharmaceutical pipeline, the findings suggest that pharmaceutical development does not adequately address global disease burden. Future research on the under-reported details of Phase I and II clinical trials is needed to understand how the industry operates and how its resource-allocation matches global health concerns [13].

1.1.2.2 Value addition through BDP table—industry targets, global burden of diseases

There is a realistic possibility for Industry through the approach of Reverse Pharmacology techniques to develop effective Botanical Drugs for some of the leading causes of mortality and still be commercially successful in the market. There is a substantial history of therapeutic use of Botanical Drug Substances (in turn to be developed as Botanical Drug Products) given in Traditional System of Medicine e.g., Indian System of Medicine for top mortality disease conditions.

Table 1.3 Botanical Drug Substances from Ayurveda, Indian System of Medicine for Some of the Leading Causes of Mortality in the World

No.	Mortality Factors in 2016	Multi Target Ligands through Botanical Drug Raw Material
1.	Ischemic heart disease and other cardiovascular disease	Terminalia arjuna, Commiphora mukul, Boerhavia diffusa, Acorus calamus, Allium cepa
2.	Lung cancer and COPD	Piper longum, Tinospora cordifolia, Azadiracahta indica, Canabis sativa
3.	Cerebrovascular disease	Bacopa moniera, Glycirrhiza glabra
4.	Alzheimer's disease	Bacopa moniera, Centella asiatica, Curcuma aromatic, Hibiscus rosa sinensis, Allium cepa
5.	Colorectal cancer	Andrographis paniculata, Eclipta alba, Curcuma aramatica, Swertia chiraita
6.	Lower respiratory infections	Piper longum, Azadiracahta indica
7.	Breast cancer	Caesalpinia crista, Asparagus racemosus
8.	Diabetes	Tinospora cordifolia, Cinnamonum zeylanicum
9.	Pancreatic cancer	Aegle marmelos, Canabis sativa, Ficus benghalensis
10.	Chronic kidney disease	Boerhavia diffusa, Tribulus terrestris
11.	Stomach cancer	Glycirrhiza glabra, Asparagus racemosus, Emblica officinalis
12.	Liver cancer	Eclipta alba, Tinospora cordifolia, Phyllanthus neruri, Rubia cordifolia
13.	Drug use disorders	Achyranthus aspera, Rubia cordifolia, Ferula narthex, Acorus calamus, Ficus religiosa
14.	Leukemia	Malaxis acuminata, Microstylis muscifera, Polygonatum verticillatum, Polygonatum cirrhifolium, Roscoea procera, Lilium polphyllum, Habenaria edgeworthii, Habenaria intermedia
15.	Prostate cancer	Allium cepa, Tribulus terrestris, Semecarpus anacardium

Botanical Drug Substances can be converted further into various other polymorphs to target disease variations by using Pharmacogenomics techniques (Table 1.3).

1.2 Economics of drug discovery

1.2.1 Current challenges in drug development

Developing new medicines is becoming an increasingly expensive business, although precisely how expensive is the subject of debate as per subjective perspective. In 2016, the Tufts Center for the Study of Drug Development estimated the average cost of developing a new drug molecule to be $2.6 billion [14]. A recent assessment by Deloitte, from the returns obtained by twelve leading biopharma companies, puts the average cost of development, from initial discovery work to regulatory licensure, at $1,539 million [15].

1.2.1.1 Key components of R&D cost

There are four determining main variables of the capitalized cost of a new drug estimate: Actual Expenditure, Profit Probability, Progress Time, and the Cost of Capital. By considering the long timeline required to develop a new drug and other associated risks, industry should asses drug failures and the cost of capital to compute in the total cost of a new successful drug, i.e., not just actual drug development expenditure but also cost of opportunity. Capitalized cost is the standard accounting treatment for long-term investments. This recognizes the fact that investors require a return on research that reflects alternative potential uses of their investment [16].

Actual expenditure: This cost has constantly increased over last 20 years due to inflation rates and costs paid towards hiring skills and employing resources. The most recent estimates are similar for out-of-pocket development costs (from Phase I through III) at around U.S. $215–220m in 2011; however, the studies are less consistent in their estimates of the magnitude of the cost of the different clinical trial phases [17–19].

Profit probability: The most recent estimates of probability of success for Phase I, Phase II, and Phase III are in between 49% and 75%, 30% and 48%, and 50% and 71%, respectively. Overall, the cumulative clinical success rate appears to have decreased over time. There is significant decrease in success rates, especially in Phase II and Phase III [20].

Progress period: Overall development time (Phases I–III) appears to have remained relatively constant over time, at around 6.5 years (75–79 months) on average. Phase III trials tend to be the longest development phase, although the most recent work suggests that development times for Phases II and III now are similar [16].

Cost of capital: The long timelines of pharmaceutical R&D mean that the cost of capital has a major impact on the final cost per successful New Molecular Entity (NME). The estimated cost per successful drug is highly sensitive to the cost of capital applied, which is now 11% up from previous figure of 9% [17].

The United States Food and Drug Administration (FDA) defines NME as a drug that contains an active moiety that has never been approved by the FDA or marketed in the United States, which means it is completely novel for the U.S. market landscape. Pharmaceutical and Biotech industry as well as market at glance perceive NME as pragmatic quantification of R&D accomplishment [20].

The pharmaceutical industry has been confronting lots of challenges in new drug discovery in order to keep up the pace with successful market launching of new drugs and earn sustainable profits out of it. If the number of new drugs approved is being evaluated in last 20 years, the conclusion is conspicuous that R&D projects are not delivering as per expectation and return on investment (ROI) on R&D investment is significantly doubtful [21]. In last

15 years, there are realistic improvements in new understanding of molecular biology, biochemistry, genetics, and new computation techniques in screening of prospective new drug compounds. Even if we expand the domain of Term Innovation in virtue of its capacity to generate new data and multiple conceptual diversity based novel therapeutic measures based thereof, pharmaceutical R&D seems to be still in a regressive mode [22].

Over the last 10 years, there are many consolidations and mergers and acquisitions are seen in among major pharma and biotech companies in order to pull resources and intellectual capital together and create exponential value however, this strategy has not offered any new innovation outcome macroscopically in the market. So overall, new drug discovery and development models in last 20 years seem to be not disruptive, and efficiency level in the system is unchanged except new data generation [23].

Target diseases on which industry is counting on and its pathophysiology and culminating factors are not completely comprehended and also human biology is itself not linear so it poses critical challenges for toxicological, pharmacological, and clinical development. The regulatory environment is also evolving rapidly since last two decades so any novel drug coming out of the pipeline has to exhibit safety and efficacy in wider-scale and longer studies. The impact of regulatory requirements on the productivity of pharmaceutical R&D suggests that this is not the most critical variable in explaining low levels of productivity [24].

Drug discovery being a complex network of critical control points possess foundational and hierarchical pitfalls. Almost close to 90% of the candidate drugs that enter Phase I trials fail. In fact, most of these drug candidates will not even enter in Phase II trial, the point at which the efficacy of a candidate drug is evaluated for the first time in humans [25].

The drug candidate's failure entering in early stage of drug development clearly demonstrates ambiguous logic in cause and effect establishment of disease pathogenesis and selection of drug candidates. This failure also constitutes evidence of pharmaceutical and biotech companies' organizational cultures in mitigating product development risks by enrolling astronomical numbers of drug candidates and set in place of stage-wise progression of drug development projects [26].

The high attrition rate for candidate drugs is at the heart of the dramatic increase in the cost of drug discovery, and this represents at least one and half times increase in inflation-adjusted terms over the previous decade. Less visible but equally important are the costs incurred by the patients enrolled in clinical trials, who are exposed to drug candidates with a high probability of failure. This direct human impact is never accounted for in conventional measures of "productivity," yet it represents a crucial dimension of the social harm caused by the closed model of drug discovery.

In any given disease, pharmaceutical R&D generally focuses on a narrow range of targets, with many organizations—in academia and industry—often pursuing the same underlying hypotheses. The case of human protein kinases is a good example. These are proteins that regulate key aspects of cellular signaling and whose modulation has provided the mechanism of action for a multitude of drugs across therapeutic areas. Yet only a small proportion of kinases have been explored so far: pharmaceutical research in both academia and industry has concentrated on fewer than 50 of the more than 500 known human kinases. Barriers to sharing information about which targets have already been evaluated, and a generalized unwillingness to explore riskier areas of the kinome, explain this concentration of effort on a very limited set of targets. As a result, large sections of the human kinome remain effectively unexplored [27,28].

Simply doubling down on traditional economic incentives will not solve these systemic weaknesses in the pharmaceutical R&D ecosystem. The history of drug discovery over the last two decades indicates that increasing the volume of research funding or extending intellectual property (IP) protections fails to enhance the overall efficiency of pharmaceutical R&D. This is an inelastic system, unable to extract the full benefit from new economic or scientific inputs, and a system that cannot be re-energized simply by intensifying conventional financial or proprietary incentives.

1.2.2 Productivity failure

There are two broad categories of productivity failure as described below, neither of which can be successfully tackled by simply reinforcing conventional funding models or IP acquisition instruments.

1.2.2.1 Complexity in disorders

Complex Disorders are unique disorders where internal and external etiological factors are involved. Neurodegenerative Diseases, Chronic Systemic Inflammation, Psychiatric Disorders, and Cancers are some of the examples of such diseases. Genetic predisposition, Lifestyle, Diet, and Environmental Factors are cumulatively responsible for such disease category. Despite the advances on the comprehension of the biological basis of these conditions and the huge investments made by the pharmaceutical sector, pharmaceutical solutions remain uncertain. As per their possible multiparameter-based etiological pathway it is logical that approach to manage and or cure such diseases mandate multi target ligands. The clinical outcome would be even stronger if techniques like predictive ADME studies, drug-drug interactions predictions are considered earlier on. There are several reviews available covering the potential of the multi-target approach in cancer [29].

1.2.2.2 Drug resistance in certain diseases

Drug Resistance in certain diseases are due to intrinsic or induced variability in drug response due to modifications in key disease-relevant biological pathways and activation of compensatory mechanisms e.g., infectious diseases.

The following are some key facts described by WHO on Drug Resistance in Infectious Diseases:

- Antimicrobial resistance (AMR) threatens the effective prevention and treatment of an ever-increasing range of infections caused by bacteria, parasites, viruses, and fungi.
- AMR is an increasingly serious threat to global public health that requires action across all government sectors and society.
- Without effective antibiotics, the success of major surgery and cancer chemotherapy would be compromised.
- The cost of healthcare for patients with resistant infections is higher than care for patients with non-resistant infections due to longer duration of illness, additional tests, and use of more-expensive drugs.

In 2016, 490,000 people developed multidrug resistant TB globally, and drug resistance is starting to complicate the fight against HIV and malaria, as well [30].

Apart from the obvious applications in the field of antimicrobial chemotherapy (it is less probable to develop resistance linked to single-point mutations against multi-target than single-target agents), this strategy could also be pertinent to treat non-infectious conditions characterized by high incidence of the drug resistance phenomena, e.g., epilepsy. One-third of epileptic patients suffer from refractory epilepsy. One of the prevalent hypotheses to explain refractory epilepsy cases proposes that at least part of the non-responsive patients might express variations in molecular targets of antiepileptic drugs [31,32]. Isobolographic studies in animal models and clinical experience suggest that combination of drugs with different mechanisms tends to be beneficial [33–35]. On the other hand, whereas there exists consensus regarding the utility of single-target drugs for the treatment of some specific epilepsy types or syndromes, broad-spectrum antiepileptic drugs such as valproic acid are among the most-used antiepileptic agents and might be valuable in those cases where, at the onset of epilepsy, diagnosis of the specific syndrome is elusive [36]. Nature has evolved distinct strategies to modulate biological processes, either by selectively targeting biological macromolecules or by creating molecular promiscuity or polypharmacology (one molecule binds to different targets). Widely claimed to be superior versus monosubstances, mixtures of bioactive compounds in botanical drugs allegedly exert synergistic therapeutic effects and in such typical disease conditions Botanical Drug Products could be beneficial in integration with conventional drugs due to its very own nature being multi molecular or exhibiting polypharmacological actions [37].

1.3 Novel drug discovery models

Drug Discovery and Development is an ideal case of the Integrated Operating Model where all project components are linearly and exponentially interrelated. Each component has its own contribution limits; however, its influencing capacity on another components would be more impactful than its natural capacity. For example, after evaluating ADME studies, decision point in determining primary and secondary therapeutic end points will have significant influence on budgeting of Phase II and Phase III trials.

Notwithstanding revolutionary early drug discovery models in last twenty-odd years the degree of serendipity is multifold increased as the data generation is happening earlier than experiments.

In the past, Drug Discovery and Development was an innovation undertaking of a single Pharmaceutical Institute where in, starting from an Idea Conception to successful Market Launch. At this point of time however, innovation is more disseminated in terms of place diversity e.g., proof of new drug concept can be from Start Up and or University Incubator, inputs on target market from Market Research Groups etc. This transformation has put burden of success not only on one specific institution rather than on various organizations.

In such disseminated Drug Innovation Model, the value preposition of shared drug discovery is obvious that it can establish greater connectivity among the parts and in so doing long-term efficiency would certainly arrive. A critical contribution of shared innovation approach is to facilitate the early sharing of information and knowledge in the riskiest stages of the drug discovery process, when biological targets for therapeutic interventions are selected or the range of relevant chemical matter is being prospected [38]. Such R&D collaboration in a shared innovation approach would terminate duplication of resource utilization and reinforcing drug development pathway in terms expanding its resources utilization.

This strategy will exert change on a pharmaceutical organization's behavior and culture as well as its IP portfolio. Targeting IP as the prime output of severely executed R&D projects is the sole cause of massive duplication of scientific experiments. But such scientific experiments duplication can drain time and efforts without any new outcome. The cost of opportunity won't be that high if it would have obtained from shared knowledge platform. Such shared knowledge platform derived information will allow pharmaceutical organizations possessing bigger resources pool to accept new experimental collaboration and conclude on more distinguished IP and market advantage. The incentives of guarding conventional IP can be enhanced with a more open-based science approach and explore newer targets and pathways definition of diseases. It facilitates the sort of collaboration across the academia-industry divide that is essential to putting drug discovery on an even stronger scientific basis.

It helps create further transparency across the system and thus removes some of the informational asymmetries that explain why huge amounts of resources are invested in projects that are doomed to fail. It expands and nourishes the drug discovery ecosystem by creating a playing field where for-profit and non-profit actors can join efforts while pursuing different strategies [39].

1.3.1 Disease morphology

The Drug Discovery & Development Project starts from the precise target of pathogenesis and how ligand target binding will behave in given biological space. That's why it is of grave importance of high or low probability of success. The validation of target and defining its boundaries in relation to desired therapeutic effect is actually risk management. In the past precisely selected target-based modulation approaches have offered therapeutically successful drugs. The incubators, university labs, startups, and consortiums are the best bet in this regard to obtain the most probable drug targets.

The target-centric approaches are desirable because these complement well with high-throughput screens and give meaningful structure–activity correlations in a lead optimization phase. However, there are shortcomings of the approach as well. The inadequate efficacy and/or non-optimal safety have emerged as limitations during clinical investigation in some instances. The binding or interaction of a potential drug molecule on a target may trigger complex response mediated through interconnected network biology, which may compensate, enhance, or negate the initial event. But these shortcomings can be detected early, modulate and or terminated right after toxicity studies are well performed through surrogate clinical trials as a proof of concept. The most of the chronic and or lifestyle disorders are multi-elemental in nature. The attunement of any given single drug target in given cellular network can influence on its therapeutic outcome. The analysis of tissue and plasma samples from disease population can help validate target hypothesis. Disease Genotyping and Phenotyping can overcome this issue in partial manner and conjugation of any other known molecule can be sought to define further specificity of drug target [19,40,41].

The most of chronic non-communicable diseases are in constant dynamic mode and drug targets can slightly deviate than predetermined level. The advantage of Phenotype Screening is to generate specificity in previously unknown serendipitous disease biology. However, low throughput and lead optimization difficulties in generating good structure–activity relationships create challenges. Both these approaches have their merits and should be investigated in preclinical exploratory phase to compare and narrow down the preferred choices.

1.3.2 Animal model translation

Animal Models depicting precise disease prototype and evaluation of drug candidates based there on is utmost critical. In the case of multifactorial diseases early drug development becomes even more critical as there are several etiological trigger factors in the disease causeway. The animal models pertinent to specific mechanism of action in such cases may need to be developed to validate hypothesis in a very early preclinical phase [42]. This is not the case in communicable or infectious diseases; animal models can translate well in further research just due to factor of single target and ligand to consider. Chronic Diseases animal model translational pitfall can be overcome through combinations of Evidence Based Medicine Model or computational algorithms. The structure-activity relationship with the most known biomarkers can be further ascertained through microarrays. The advantage with Botanical Drug Products is that non-clinical requirements are limited or reduced in investigational new drug (IND) filing as previous human experience can fulfill the gap of proof of safety.

1.3.3 Toxicity

The risk of Idiosyncrasy or Toxicity can be mitigated if in-vivo and in-vitro safety analysis and geno-toxicity and muta- genicity studies are performed in animal multi species. This step will further reinforce in qualifying drug candidates approaching bedside. This step is crucial in terms of long-term treatments that is usually mandated for chronic disorders, e.g., neurode-generative disorders, blood oncological conditions, and cardiovascular and kidney disorders. The regulatory approval advantage in Botanical Drug Product is that, FDA evaluates prospective drug candidate Toxicity studies on a case-to-case basis, like if it was previously marketed in any other form and for how much time. Also the selection of appropriate biomarkers or surrogate endpoints that can build confidence on safety of target or rule out any "off-target" effects observed in preclinical screens can help elimi-nate less promising candidates early [42].

1.3.4 Therapeutic quality of selected drug candidate

For business organizations such as Start Ups and Incubator-based Set Ups, the prospective drug candidate once determined and progressed little bit in drug discovery pathway has to proceed further until its fate determination point arrives. In case it gets dropped can falsify target assumption and further drug development gets early terminated and can incur significant economic loss. That's why such organizations start with multiple exploratory experiments and terminate some if predetermined criteria aren't accomplished. In the case of Botanical Drug Products, the genomics and bioinformatics can help com-pound discovery from an unknown pursuit to a high-throughput possibility. The algorithms to mine metagenomic data and identification biosynthetic gene clusters in genome sequences can predict the chemical structures of multiple compounds in given single poly-molecular Botanical Drugs. Few network-ing strategies can then incorporate to systematize large volumes of genetic and chemical data and establish that information to metabolomic and phe-notypic data. Further the efficiency of process and quality of lead generated can be improved by judicious selection of available technologies at various stages of drug discovery such as target ID and/or validation (over expression and knockout), hit generation phase (X-ray crystallography), structure-guided drug discovery (SGDD), fragment-based, virtual screening, high-throughput screening (HTS) to lead optimization (scaffold hopping, allosteric versus active site modulation, drug pharmaco-kinetics properties such as absorption, distribution, metabolism and excretion (ADME), selectivity and safety screens [43–45].

Early Phase projects that of Lead Optimization must have a critical stop point determination in order to validate and reinforce compound profiling in terms of clinical bedside competition and advantage. This could be true test of innovation and its value preposition. The quality of molecules can be determined based on their pharmacokinetic and/or pharmacodynamics correlation, ADMET profile, and good therapeutic index.

1.3.5 Innovative clinical investigation

Later in the Drug Development Pathway the clinical trial models must be optimized as such that it should target to evaluate critical efficacy and safety as early as possible in a cost-effective manner. In order to achieve this goal, one can employ Clinical Phenotyping, in-silico simulation, real time bio-markers calibration, structural imaging, qualitative scoring systems and or surrogate clinical endpoints to evaluate the market potential of the drug molecule. In addition to the techniques discussed, the microdosing (Phase 0) can be useful in select cases to prioritize compounds and get an early assessment of human pharmacokinetics profile. Using a microdose of a radiolabeled investigational drug at 1/100th predicted therapeutic dose, in combination with positron emission tomography (PET), can help get an early assessment of human pharmacokinetics profile and targeted tissue distribution (brain, tumor, among others). The microdosing approach assumes linearity in the pharmacokinetics profile during dose scaling but can yield useful early information in a cost-effective manner with low compound needs [42].

1.4 Strategic positioning of botanical derived molecules, phytochemicals in health care market

Phytochemicals derived from medicinal plants constitute a significant percentage of drugs in the therapeutic areas in communicable and non-communicable diseases. The conventional but proven way earlier was single compound/ single target and or receptor approach so that maximum therapeutic sensitivity can be achieved in minimalistic side effects [46]. As described earlier in this chapter this approach is proven to be no more practicable in the market as considerable drop in effective drugs coming from bedside to bench side. The significant contribution in this innovation deficit is a post-market failure of blockbuster drugs. Many analysts believe that the current capital-intensive model—the one-drug-to-fit-all approach—will be unsustainable in future and that a new less-investment, more-drugs model is necessary for further scientific growth [46].

Complex disorders are multifactorial in nature and there are more than pathogenesis pathways involved in manifestation of disease. Also these n pathways trigger several other alternative pathways in response to specific target inhibition and exhibit resistive modulation due to which drug candidate likely fails in Clinical Trial Test.

To manage complex diseases like Connective Tissue Disorders, Metabolic Disorders, Oncological Diseases solitary pathogenesis and or target can't be aimed at to develop a single drug rather it has to be in a league Polypharmacology form to mitigate risk of clinical failure and drug resistance. Integrating network biology and poly-pharmacology holds the promise of expanding the current concepts on druggable targets [47].

The dominant paradigm in drug discovery is the concept of designing maximally selective ligands to act on individual drug targets. In Botanical Drug Products, molecules are connected to each other in an ordered manner defined by binding interactions in time and space rather than in a loose assemblage [37]. The different interactions between various components might involve the protection of an active substance from decomposition by enzymes, modification of transport across membranes of cells or organelles, evasion of multi-drug resistance mechanisms among others.

To overcome the clinical failure, therapeutic drug resistance, and severe side effects of conventional drugs in complex disorders, Botanical Drug Products could be an effective way out. As botanical Drug Products has several synergistic molecules in action, using network pharmacology optimization in accord with metabolomics pathway of disease can generate superior single drug than combination therapy [46].

1.5 Conclusion

Industrial Pharmaceutical Innovation in past decades is not delivering at the expecting level of resources utilized in it, so ROI in big-ticket-size projects is questionable, however industry is still investing in escalated manner in a hope to have major breakthrough in understanding of complex diseases and potential targets to intervene. The majority of the global population is living longer than the previous 75-years average life span; therefore, there is an effect in increased number of complex disorders, chronic disorders, and poor quality of life. As no new clinically better healthcare interventions are coming in clinics, the overall healthcare quality is also diminishing at the consumer end, hampering prosperity and confidence. Industry has to adopt new drug discovery techniques, integrated programs and blue-sky avenues like Botanical Drug Products. Botanical Drug Products qualified drug discovery and development leads are contained in a sustainable ethnic healthcare system such as Indian System of Medicine and Traditional Chinese Medicine.

References

1. Pharma 2020: From vision to decision. Available at https://www.pwc.com/gx/en/industries/pharmaceuticals-life-sciences/publications/pharma-2020.html.
2. United Nations, World population prospects: The 2010 revision, volume I: Comprehensive tables. ST/ESA/SER.A/313, Department of Economic and Social Affairs, Population Division, 2011.
3. World Health Organization, *Global Status Report on Noncommunicable Diseases 2010*, World Health Organization, Geneva, Switzerland, 2011.
4. World Health Organization and Public Health Agency of Canada, eds., *Preventing Chronic Diseases: A Vital Investment*, World Health Organization; Public Health Agency of Canada, Geneva, Switzerland [Ottawa], 2005.
5. P. Martin, *The Prevalence of Dementia Worldwide*, Alzheimer's Disease International, 2008.
6. QuintilesIMS, *Outlook for Global Medicines through 2021*, QuintilesIMS Institute, 2017.
7. Evaluate Pharma, *World Preview 2016, Outlook to 2022*, Evaluate Pharma, London, UK, 2016.
8. National Institutes of Health (NIH). Budget. Available at https://www.nih.gov/about-nih/what-we-do/budget.
9. B. Kyle, K. Mosahid, L. Ryan, T. Gerard, Y. Yukio and Z. Hao, *Patent Cooperation Treaty Yearly Review: The International Patent System*, WIPO, 2016.
10. M. Naghavi, A.A. Abajobir, C. Abbafati, K.M. Abbas, F. Abd-Allah, S.F. Abera et al., Global, regional, and national age-sex specific mortality for 264 causes of death, 1980–2016: A systematic analysis for the Global Burden of Disease Study 2016, *The Lancet* 390 (2017), pp. 1151–1210.
11. IFPMA, *IFPMA Facts and Figures Report: IFPMA*, IFPMA, 2017.
12. Evaluate Pharma, *World Preview 2015, Outlook to 2020*, Evaluate Pharma, 2015.
13. M.D. Cottingham, C.A. Kalbaugh and J.A. Fisher, Tracking the pharmaceutical pipeline: Clinical trials and global disease burden, *Clin. Transl. Sci.* 7 (2014), pp. 297–299.
14. J.A. DiMasi, H.G. Grabowski and R.W. Hansen, Innovation in the pharmaceutical industry: New estimates of R&D costs, *J. Health Econ.* 47 (2016), pp. 20–33.
15. Deloitte Center for Health Solutions, *Balancing the R&D Equation: Measuring the Return from Pharmaceutical Innovation 2016*, Deloitte Center for Health Solutions, New York, 2016.
16. J. Mestre-Ferrandiz, J. Sussex, A. Towse and Office of Health Economics, *The R&D Cost of a New Medicine*, Office of Health Economics, London, UK, 2012.
17. J.A. DiMasi, R.W. Hansen, H.G. Grabowski and L. Lasagna, Cost of innovation in the pharmaceutical industry, *J. Health Econ.* 10 (1991), pp. 107–142.
18. C.P. Adams and V.V. Brantner, Spending on new drug development, *Health Econ.* 19 (2010), pp. 130–141.
19. S.M. Paul, D.S. Mytelka, C.T. Dunwiddie, C.C. Persinger, B.H. Munos, S.R. Lindborg et al., How to improve R&D productivity: The pharmaceutical industry's grand challenge, *Nat. Rev. Drug Discov.* 9 (2010), pp. 203–214.
20. F. Pammolli, L. Magazzini and M. Riccaboni, The productivity crisis in pharmaceutical R&D, *Nat. Rev. Drug Discov.* 10 (2011), pp. 428–438.
21. B. Munos, Lessons from 60 years of pharmaceutical innovation, *Nat. Rev. Drug Discov.* 8 (2009), pp. 959–968.
22. R&D productivity: On the comeback trail. Available at https://www.nature.com/articles/nrd4320.
23. P. Nightingale and P. Martin, The myth of the biotech revolution, *Trends Biotechnol.* 22 (2004), pp. 564–569.
24. B. Booth and R. Zemmel, Prospects for productivity, *Nat. Rev. Drug Discov.* 3 (2004), pp. 451–456.

25. K. Smietana, M. Siatkowski and M. Møller, Trends in clinical success rates, *Nat. Rev. Drug Discov.* 15 (2016), pp. 379–380.
26. D. Cook, D. Brown, R. Alexander, R. March, P. Morgan, G. Satterthwaite et al., Lessons learned from the fate of AstraZeneca's drug pipeline: A five-dimensional framework, *Nat. Rev. Drug Discov.* 13 (2014), pp. 419–431.
27. The (un)targeted cancer kinome. Available at https://www.nature.com/articles/nchembio.297.
28. S. Knapp, P. Arruda, J. Blagg, S. Burley, D.H. Drewry, A. Edwards et al., A public-private partnership to unlock the untargeted kinome, *Nat. Chem. Biol.* 9 (2013), pp. 3–6.
29. A. Talevi, Multi-target pharmacology: Possibilities and limitations of the "skeleton key approach" from a medicinal chemist perspective, *Front. Pharmacol.* 6 (2015).
30. Antimicrobial resistance. Available at http://www.who.int/news-room/fact-sheets/detail/antimicrobial-resistance.
31. M.T. Bianchi, J. Pathmanathan and S.S. Cash, From ion channels to complex networks: Magic bullet versus magic shotgun approaches to anticonvulsant pharmacotherapy, *Med. Hypotheses* 72 (2009), pp. 297–305.
32. M. Emiliano Di Ianni and A. Talevi, How can network-pharmacology contribute to antiepileptic drug development? *Mol. Cell. Epilepsy* 1 (2014), e30.
33. P. Kwan and M.J. Brodie, Combination therapy in epilepsy: When and what to use, *Drugs* 66 (2006), pp. 1817–1829.
34. R.M. Kaminski, A. Matagne, P.N. Patsalos and H. Klitgaard, Benefit of combination therapy in epilepsy: A review of the preclinical evidence with levetiracetam, *Epilepsia* 50 (2009), pp. 387–397.
35. J.W. Lee and B. Dworetzky, Rational polytherapy with antiepileptic drugs, *Pharm. Basel Switz.* 3 (2010), pp. 2362–2379.
36. L. Lagae, The need for broad spectrum and safe anti-epileptic drugs in childhood epilepsy, *Eur. Neurol. Rev.* (2006), pp. 36.
37. J. Gertsch, Botanical drugs, synergy, and network pharmacology: Forth and back to intelligent mixtures, *Planta Med.* 77 (2011), pp. 1086–1098.
38. M.J. Waring, J. Arrowsmith, A.R. Leach, P.D. Leeson, S. Mandrell, R.M. Owen et al., An analysis of the attrition of drug candidates from four major pharmaceutical companies, *Nat. Rev. Drug Discov.* 14 (2015), pp. 475–486.
39. C. Bountra, W. Lee and J. Lezaun, *A New Pharmaceutical Commons: Transforming Drug Discovery*, University of Oxford, Oxford, UK, 2017.
40. D.C. Swinney and J. Anthony, How were new medicines discovered? *Nat. Rev. Drug Discov.* 10 (2011), pp. 507–519.
41. M. Rask-Andersen, M.S. Almén and H.B. Schiöth, Trends in the exploitation of novel drug targets, *Nat. Rev. Drug Discov.* 10 (2011), pp. 579–590.
42. I. Khanna, Drug discovery in pharmaceutical industry: Productivity challenges and trends, *Drug Discov. Today* 17 (2012), pp. 1088–1102.
43. S. Andersson, A. Armstrong, A. Björe, S. Bowker, S. Chapman, R. Davies et al., Making medicinal chemistry more effective: Application of Lean Sigma to improve processes, speed and quality, *Drug Discov. Today* 14 (2009), pp. 598–604.
44. N.A. Meanwell, Synopsis of some recent tactical application of bioisosteres in drug design, *J. Med. Chem.* 54 (2011), pp. 2529–2591.
45. M.P. Gleeson, Generation of a set of simple, interpretable ADMET rules of thumb, *J. Med. Chem.* 51 (2008), pp. 817–834.
46. M.A. Rather, B.A. Bhat and M.A. Qurishi, Multicomponent phytotherapeutic approach gaining momentum: Is the "one drug to fit all" model breaking down? *Phytomedicine Int. J. Phytother. Phytopharm.* 21 (2013), pp. 1–14.
47. A.L. Hopkins, Network pharmacology: The next paradigm in drug discovery, *Nat. Chem. Biol.* 4 (2008), pp. 682–690.

2

Regulatory Perspectives of Botanical Drug Products in Top Pharmaceutical Markets

Grace Checo, Jayant N. Lokhande, and Sonali Lokhande

Contents

Botanical products have increased in use over the past three decades as a form of complementary and alternative medicine [1]. Patients around the world are attracted to their use as a means for natural treatment, falsely assuming they are safe and free of side effects [2]. Though they are not as chemically altered as the more-common synthetic drugs, which can often induce harsh side effects ranging from the breakdown of muscle tissues to even death, botanical products still have the potential to adversely affect consumers. Plants, being products of nature as opposed to laboratory synthesis, are not wholly understood; therefore, when botanicals are used as sources of medicinal compounds, their active ingredients cannot always be identified [2]. As a result, absolute explanations for how their consumption may affect physiological function or how their constituents may react with other drugs can remain unknown [1]. Without clear evidence that a botanical product will be effective as treatment, it cannot be marketed as a drug for sale or be prescribed by healthcare professionals.

With an increased interest in botanical products, countries have begun to create regulations for botanical drug development and marketing. These regulations are created to closely resemble the procedures for developing and marketing synthetic drugs. According to the World Health Organization (WHO), too few botanicals had known safety and efficacy information, so in 1998 WHO published guidelines to assist international drug authorities and manufacturers with developing botanical drugs products that promise safe and effective use [3]. However, the complexity of botanical compounds makes the procedures challenging to follow. Luckily, some botanicals have been used for so long as natural remedies that they have become part of tradition. Products that utilize these traditional compounds may be permitted a shortcut through the regulated procedure because their tradition substantiates their therapeutic claim. For all the other products, the procedure is much longer. At the heart of these regulations is ensuring the health and safety of patients. Most regulations follow a similar procedure of requiring data to confirm the safety and efficacy of the botanical product. The regulations in different countries vary in the amount of data required, and the source of the data obtained. Nevertheless, once approved by the government agency, botanical drugs are placed on the market and continually monitored for unanticipated side effects. Any botanical drugs that are deemed too hazardous even after government approval have the potential to be removed from the market.

The first botanical drug product (BDP) to receive approval by the United States Food and Drug Administration (FDA) is Veregen™, a topical ointment used to treat external genital and perianal warts caused by the human papillomavirus (HPV) in adults. Veregen was not given FDA approval until late 2006; 2 years after the initial FDA BDP regulations were issued. The drug substance in Veregen is Kunecatechine, which comes from the water extract of the green tea leaf *Camellia sinensis (L.) O Kuntzel*. Eighty-five to ninety-five percent of its weight comes from 8 known catechins, and 2.5% of its weight comes from gallic acid, caffeine, and theobromine. The remaining percent of the drug substance's weight, however, is derived from undefined botanical substituents of the green tea leaves. Consequently, it is unknown if these associated undefined compounds contribute in any way to the treatment of HPV warts [4]. The second ever BDP to receive FDA approval is Mytesi® (Fulyzaq®), a delayed-release tablet used to treat non-infectious diarrhea in adult HIV/AIDS patients caused by anti-retroviral therapy. FDA approval was given in 2012, 6 years after the first BDP described above, and 8 years after the issuance of the BDP Guidance for Industry. The botanical drug substance in Fulyzaq is crofelemer, which is extracted from the red sap of *Croton lechleri,* often referred to as the dragon's blood tree in South America [5]. Though not all components of these drugs were identified, their safety and efficacy were validated over a long procedure, and they gained passage to the U.S. market.

2.1 USA

Botanical products can exist as foods (dietary supplements), medical devices (gutta-percha), or cosmetics, depending on their intended use; however, this chapter will focus specifically on those that exist as drugs. As outlined in 2004 by the Center for Drug Evaluation and Research (CDER) under the U.S. FDA, a *botanical drug* is defined as follows:

> *A botanical drug product is intended for use in the diagnosis, cure, mitigation, treatment or prevention of disease in humans.*

- *A botanical drug product consists of vegetable materials, which may include plant materials, algae, macroscopic fungi, or combinations thereof.*
- *A botanical drug product may be available as (but not limited to) a solution (e.g., tea), powder, tablet, capsule, elixir, topical, or injection.*
- *Botanical drug products often have unique features, for example, complex mixtures, lack of a distinct active ingredient, and substantial prior human use. Fermentation products and highly purified or chemically modified botanical substances are not considered botanical drug products.*

The development of botanical drug products is regulated by the FDA within the U.S. Department of Health and Human Services. To become a legitimate drug that is prescribed by healthcare professionals, there is a long process that must be completed to ensure the safety and efficacy of the new drug. A sponsor is the pharmaceutical company or research institution that takes responsibility of conducting the proper laboratory experiments to develop the new drug, and then cooperates with the FDA to complete the rigorous evaluation process prior to official marketing. After the initial development of the botanical drug, it must undergo preclinical animal testing on multiple species to test that the botanical substituent(s) actually exhibit(s) the intended activity. Once its activity is established, toxicity must be tested in preclinical trials to determine if the drug is safe enough to begin human testing. At this point, preclinical trials are complete, and the sponsor must approach the FDA with a proposed clinical protocol to begin human trials; this stage comprises the investigational new drug (IND) application.

The IND information for a botanical drug varies depending on the category of botanical, i.e., if the botanical drug has been lawfully marketed in the United States and if the botanical product has safety concerns. If a BDP has previously been investigated or marketed in another country but then withdrawn due to extreme safety hazards, it should be noted. If the BDP has not been previously marketed at all, the location and description of the botanical's habitat must be identified, along with how and when it grows, and any risk of the species' extinction. Generally, the purpose of any IND is to provide information on the safety and efficacy of the BDP as a drug agent. It should outline manufacture control for quality consistency and explain planned procedures for clinical trials. The pharmacological effects must be included and should describe the chemical properties and structural formula of the botanical constituent, as well as the biological effects of the BDP. This section should explain, for example, the mechanism by which the drug alters physiological function and prevents disease. In some botanical drugs, all botanical substituents may come from the same part of the plant. In this case, FDA regulations state that the sponsor does not need to determine the clinical effects of each independent entity of the botanical product, which would be a challenging task because it is often difficult to distinguish the active constituent within a botanical drug. Instead, the FDA asks for an overall explanation of the BDP's effects as a whole.

Toxicological effects must also be included in the IND to describe the nature and mechanisms of deleterious effects of the botanical substance(s). Because the majority of this information is obtained through initial animal testing, the effects of BDPs in animals as determined through the results of preclinical trials should be included. If any information obtained with respect to the BDPs potential for drug dependence and abuse, it should be included in the IND. If the botanical substance has been previously tested on humans, those results should be included as well, with a reference to the corresponding INDs and a comparison between the two. Any anticipated risks and side-effects should be included in the IND and taken into consideration when designing the procedures for the anticipated clinical trials. In addition, the sponsor must include information regarding the BDPs movements to get into its functional position and exit after inside the animal's body, also known as drug disposition and pharmacokinetics, respectively. If any of this information is known about humans, it too must be included in the IND. The levels of botanical substance in each drug dose need to be equivalent, especially when considering how different dosages would produce inconsistent results in both preclinical and clinical trials. Therefore, sponsors must include in their IND a method for controlling and directly measuring the amount of drug in each batch. There should be quality controls for both the raw botanical substance and the final drug product.

If the IND application is approved by the FDA, the BDP moves on to clinical trials. For this reason, protocol design is required as part of the IND. It must specify not only the experimental procedure but also the mode by which participants are selected and/or excluded; how to determine the doses administered to each individual and the duration of exposure to the BDP for each individual must be explained as well. Each clinical trial is separated into three phases. Phase 1 involves testing the BDP in a small group of healthy volunteers—usually in the tens—to identify the most frequently occurring side effects, and to understand how the drug is metabolized and excreted. If Phase 1 is completed successfully, meaning unanticipated toxicity levels are *not* found, the clinical trial proceeds to Phase 2 where there is a larger number of volunteers—usually in the hundreds. The focus of this phase is to determine if the BDP remains effective in individuals who have certain diseases unrelated to that being treated by the BDP. Again, safety and side effects continue to be monitored. Phase 3 is an expansion of Phase 1 and 2 with the goal of acquiring the same data but with an even larger, broader population—in the thousands. The BDPs safety and efficacy is tested at different dosages and in combination with other drugs.

After the completion of clinical trials and a review meeting between the FDA and drug sponsor, an official New Drug Application (NDA) is submitted formally asking the FDA to approve a drug for marketing in the United States. This application contains all data and relevant analyses accumulated from both preclinical and clinical trials. The FDA reviews this application, and an FDA Review Team authorizes the sponsor's research on the BDPs safety and efficacy. For BDPs that are high risk, the FDA must determine if the clinical benefits outweigh any adverse effects. After the green light is received, drug labeling is reviewed by the FDA to make sure appropriate information is accessible to healthcare professionals and consumers. Then the facility of the drug manufacture is inspected, and finally, the FDA approves the BDPs NDA. The drug is finally available in the U.S. pharmaceutical market [6].

2.2 Canada

Health Canada is the government agency responsible for regulating the addition of botanical products to the Canadian pharmaceutical market. Botanical products in Canada are referred to as natural health products (NHP) and are defined in Pathway for Licensing Natural Health Products used as Traditional Medicines as follows (Natural Health Products Regulation 2003):

A natural health product is a homeopathic medicine, or a traditional medicine, that is intended to provide a pharmacological activity or other direct effect in:

- *diagnosing, treating, mitigating, or preventing a disease, disorder, or abnormal physiological state or its symptoms in humans;*
- *restoring or correcting organic functions in humans; or*
- *modifying organic functions in humans, such as modifying those functions in a manner that maintains or promotes health.*

The first step in being permitted to market a natural health product in Canada is to attain a license for both the product being sold and the site(s) from which manufacturing, labeling, distribution, and so on, is taking place. There are no fees charged to submit license applications to Health Canada [7]. The license application must be submitted to the Health Canada Minister with similar types of information found in United States IND applications. Each ingredient in the natural health product, both medicinal and non-medicinal, must be identified by common name and scientific name, with a description and location of their habitat, and a summary of how they are processed. The quantity per dosage and the level of potency of the natural substance(s), and the NHP's recommended conditions of use must be provided. Recommended conditions of use include the drug's purpose, dosage form, dosage size, route of administration, duration of use, and risk information. All non-medicinal ingredients must be listed with a clear statement identifying their purpose in the drug.

After the Minister has issued a product license to the sponsor and marketing of the drug begins, the sponsor must keep an up-to-date list reporting any adverse side effects. Then, if the Minister decides the NHP is not safe enough for distribution, the Minister will request a copy of the list and the Minister may choose to recall the NHPs marketing.

The site license, which must be attained prior to NHP marketing, must also be acquired via an application to the Minister. In this application, the address and task of all sites involved in the manufacturing, packaging, labeling, and/or importing the NHP must be identified. The equipment used to complete each task, and the process by which the task is completed must be summarized as well. Once attained, the site license expires within a year and must be renewed by the Minister every 1–3, years depending on the amount of time the site license has existed.

In Health Canada's NHP regulations there is an entire section dedicated to defining Good Manufacturing Practices within the sites subject to licensing. These regulations are meant to set quality control for each batch of the NHP, and to clearly identify sanitation requirements for the premises, equipment, and employees involved in production and distribution of the drug. These facilities are inspected by government officials of Health Canada to ensure continual, proper maintenance of facilities and responsibilities [7].

2.3 India

Botanical Drug Products, Phytochemicals, Botanical-derived Chemicals and other related products are part of Ayurveda, Indian System of Medicine.

Drugs and Cosmetics Act (DCA) in 1962 has constituted and incorporated following rules:

- Ayurveda, Siddha, Unani drugs, and their reference books were included in DCA act and Section 33B to 33-O of the Chapter IV A of the DCA pertained to ASU drugs.
- Rule 158 (B) of DCA offers the guidelines for ASU drugs in India.
- Rule 161 (B) offers guidelines on label preparation including shelf life/date of expiry of ASU.
- Rule 168 (B) s pertained to the regulation of standards for manufacture, distribution, and sale of ASU formulations to ensure their safety for human consumption.

Hallmarks of Events in regulating Botanicial Drug Products

- Dr. Nitya Anand and Dr. DBA Narayana in 2008 designated the term Phytopharmaceuticals to Botanical Drug Products and or Botanically Derived Phytochemicals. Department of AYUSH with reference to GSR 702 on October 24, 2013 published the "good clinical practice guidelines for clinical trials in Ayurveda, Siddha and Unani medicine" to be followed by industry and academia.
- ASU Drugs Technical Board was reconstituted in 2015 to advise the central and state governments on technical matters pertaining to ASU drugs.
- In 2015, phytopharmaceuticals were distinguished as New Drugs from ASU Drugs under the Drugs and Cosmetics Act according to the Gazette notification GSR 918(E) [8].

A Phytopharmaceutical Drug is a purified and standardized fraction with defined minimum four bio-active or phytochemical compounds of an extract of a medicinal plant or its part which are intended for internal or external use of human beings or animals for diagnosis, treatment, mitigation, or prevention of any disease or disorder but not for parenteral route. Generally they include processed or unprocessed standard material derived from plants or parts thereof or combination of parts of plants, extracts, or fractions thereof in a dosage form [9] (Figure 2.1).

Central Drugs Standards Control Organization (CDSCO) and Min. of Health & Family Welfare, Govt. of India department of AYUSH (A=Ayurveda,

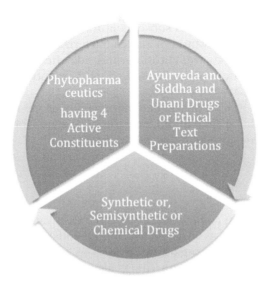

Figure 2.1 *Phytopharmaceuticals regulatory position.*

Y=Yoga and Naturopathy, U=Unani, S=Siddha and H=Homeopathy) are prime regulatory organizations regulating the phytopharmaceuticals.

Other laws governing regulation of Phytopharmaceuticals either in processing, researching and manufacturing include:

- Drug & Cosmetic Act, Chapter IV A, Rule 161, Schedule T & Schedule 1 [10].
- Drugs & Magic Remedies Objectionable Advertisement Act [11].
- Biodiversity Bill [12].
- Environment & Forest Act [13].
- Indian Patents Act [14].

Drugs & Cosmetic Rules 1945, amended in 2015 by CDSCO, has offered a strategic pathway to regulate Phytopharmaceuticals. The regulatory guidelines for Phytopharmaceuticals include scientific data on quality, safety, and efficacy and permit marketing of a plant-based drug.

Any drug lead to be registered as Phytopharmaceutical Drug with Appendix I B of Schedule Y primarily requires data on the following:

- Application to conduct clinical trial or import or manufacture
- Dosages and route of administration
- Therapeutic class
- Published literature
- Book references
- Composition
- Manufacturing process of formulation
- Identification, authentication, and source of the plant used for extraction and fractionation
- Process for extraction and subsequent fractionation and purification
- Active molecules constitution and characterization
- Research data on contraindications and side effects
- Any other on Safety per published scientific reports

After CDSCO approves NDA, the marketing status of the new phytopharmaceutical drug would be similar and have the same status for marketing as that given for a synthetic compound-based drug [15].

2.4 Australia

Herbal medicine in the complementary medicines is called therapeutic goods with herbal substances as major active ingredient(s), and they are preparations of plants and other organisms treated as plants in the International Code of Botanical Nomenclature fungi, algae, and yeast [16,17]. Botanical Drug Products or Botanically derived Drug preparations are considered under Australia's Therapeutic Goods Administration (TGA) that is responsible for regulation of therapeutic goods like medicines, medical devices, blood products, and complementary medicines including vitamins, minerals, and supplements.

2.4.1 Current designation of complementary medicines in Australia

In Australia, complementary medicines are considered as:

- Herbs
- Nutritional supplements,
- Vitamins
- Minerals
- Aromatherapy
- Homoeopathic medicines
- Traditional Chinese Medicines,
- Ayurveda
- Australian Indigenous Medicine
- Western Herbal Medicine

Therapeutic Goods Administration (TGA), supported by Therapeutic Goods Regulations (1990) and various Therapeutic Goods Orders (TGOs), regulates the complementary medicines in Australia under the Therapeutic Goods Act (1989) and offer guidelines for import, export, manufacture, and supply of therapeutic goods [18,19].

TGA uses risk-based pre-market assessment procedures in determining risk and the evaluation process to be applied to complementary medicines like registered procedure for high-risk medicines and listed procedure for low-risk medicines. A number of factors are involved in the evaluation of complementary medicines, like toxicity of the ingredients, dosage form, therapeutic claims, and adverse effects of the medicine [20,21].

1. Lower-risk medicines: Required to be listed on the Australian Register of Therapeutic Goods (ARTG), and they are not evaluated individually by the TGA. There are 12,000 complementary medicines available on the Australian market, and the majority of these medicines are listed on the ARTG. These listed medicines contain certain low-risk ingredients and must follow the Good Manufacturing Practice (GMP). These listed medicines can only make indications for non-serious indications and self-limiting conditions like health maintenance, health enhancement, and so on. A AUST L number is allocated to this class of medicine, and it must display on the medicine label. Higher-risk medicines: They must be registered on the ARTG and involve the evaluation of the quality, safety, and efficacy of the product.
2. There are 200 products registered to the ARTG, and all medicines in this class must have AUST-R marked on the label.

TGA will conduct post-market monitoring activities to evaluate the quality and safety of listed medicines including:

- Random and targeted audits of listed medicines
- Laboratory testing of complementary medicines in a manner comparable to the testing regime for over-the-counter medicines
- Random and targeted surveillance in the market-place
- Monitoring and expert review of reported adverse events
- An effective, responsive and timely recalls procedure

TGA assess the adverse events reported by consumers, health professionals, the pharmaceutical industry, international medicines regulators etc. and sponsors of medicines are required to report to the TGA suspected adverse reactions for their medicines that they are aware of.

The advertising and marketing of complementary medicines should be conducted through promoting the quality use of the product with social responsible and not to mislead or deceive the consumer. Advertising should be subjected to the advertising requirements of the Therapeutic Goods Act, adopts the Therapeutic Goods Advertising Code (TGAC), and Trade Practices Act 1975. A complaint can be lodged by anyone about an advertisement, and all those complaints will be treated confidentially.

Complementary medicines in Australia are the most strictly regulated in the world, and unfortunately recent environment of escalating red tape has contributed significantly to the difficulties faced by industry which led to stifling of product innovation, lower productivity, less job creation, minimized incentive for industry investment, and high price of operating in Australia creates a global disadvantage for local operators to compete globally.

The complementary medicines industry, the commercial reality of *no patent protection* has limited the incentive to invest in new/improved research and in this area complementary medicines are always at a disadvantage compared to pharmaceuticals [22–24].

2.5 China

Traditional Chinese Medicine (TCM) has a 4,000-year history, and its importance in ancient times was closely related to the culture, society, and art.

The evolution of TCM includes:

- Materia Medica composed in Tang Dynasty in the era of AD 618-907 by Xin Xiu Ben Cao is containing information on 844 medicinal herbs.
- Tang Shen Wei from Sung Dynasty in AD 96-1279 increased the Ben Cao work by including 1,746 substances.

- Later in AD 1590 Li Shi Zhen compiled the monumental work named Ben Cao Gang Mu, which is a compendium of Material Medica that describes 1,892 species, 11,096 herbal formulas, and nearly 40 forms of medicines.
- Current Ben Cao includes information on more than 6,000 species of medicinal plants.

In China, pharmacopoeias define the country's drug standards, and its 2005 edition includes 1,146 monographs. Currently the traditional herbs and conventional plant medicines are considered as new drugs in China. After over 2,000 years of evidence-based implementation of traditional medicine, they have showcased incredible modernized botanical drug development. The current research on Traditional Chinese Medicines isolates the active ingredients from natural products and understands their effects on body tissues by using methods like thin-layer chromatography and high-performance liquid chromatography, which are considered as acceptable from a western perspective [25–27].

Traditional Chinese Medicine is regulated as medicinal products with the same regulatory requirements as conventional medicines that require extensive pre-clinical and clinical studies. The drugs, which have not been manufactured previously in China or new indication of existing drug, change in route of administration or dosage form of existing drug are called new drugs. These new drugs will be examined and regulated as per the Drug Administration Law of the People's Republic of China. A new drug certificate with an approval number will be granted after the approval, and the applicant is permitted to put their approved product in the market [28].

The Drug Administration Law of the People's Republic of China includes:

- Article 3 and Article 5 that deal with the drug manufacturing enterprise.
- Article 11 deals with the control of drug handling enterprises.
- Articles 29 and 30 deal with the export and import of the herbal medicines
- Articles 9, 40, 45, and 71 deal with Approval numbers and Labels.
- Appendix I in 9 categories mentions classes like chemical drugs, biological products, and Traditional Chinese Medicines (natural medicines).
- SFDA in Jan 2008 released "Supplementary Regulations for Registration of Chinese Herb Products" and "Guidelines for Natural Medicinal Research" to distinguish between modern herbal medicines and traditional herbal medicines [27].

Article 21 in conjunction with Article 22 of the Drug Administration Law of 1985 of the Drug Administration Law, Ministry of Public Health, or the health bureau of the province provides clinical trial or clinical verification of a new drug.

Table 2.1 Categories and Respective Product Inclusions

Category	Description
I	• Newly discovered medicinal plants and their preparations • Active principle extracted from TCM plants material and their preparations • Artificial imitations of TCM herbs
II	• Newly employed as a remedy • Medicinal herbal injections • Non-single components extracted from TCM • TCM materials obtained by artificial techniques in vivo and their preparations
III	• New TCM preparations • Combined preparations of TCM and modern medicine
IV	• Cultivated TCM material • New dosage forms of TCM drug
V	• New use of TCM

New Traditional Chinese Medicines are classified in five categories (see Table 2.1)

The major difference between clinical trial and clinical verification is clinical trials shall be conducted on categories 1, 2, and 3, and clinical verifications shall be conducted on categories 4 and 5 of new medicines.

Applicants should provide data on toxicity, pharmacological properties, and clinical research as well as a detailed documentation on the quality of the medicinal material and the pharmaceutical form. Different requirements are essential for the above five categories to fulfill the medicinal material and pharmaceutical preparation requirements. Proprietary medicines and new medicines are exempted from the clinical testing when there is only the change in dosage form without any changes in the indications for cardinal symptoms or dosage.

The following items should be provided in the medicinal material application for clinical research like purpose of research, previous human experience, source of medicinal material, cultivation, processing, properties, information based on Chinese pharmacology and experience, efficacy, pharmacology, acute toxicity, mutagenicity, carcinogenicity, reproductive toxicity (only for category 1), quality standards, stability, and the proposed plan. Information on quality standards, stability, summary of clinical studies, and packaging material should be provided for the separate application of production.

All the reports should meet the similar requirements medicinal material reports depending on the drug category. All the technical requirements required for the pharmacological studies were laid down in the special monograph and the tests involved in the major drug effects should be designed that special characteristics of the Traditional Chinese Medicine are taken into consideration.

Based on the effects of the new medicine on the complex of symptoms or the illness, two or more methods should be selected for research on the major drug actions.

Research should be sufficient to verify the therapeutic functions and other important therapeutic effects in case of new medicines in category 1, 2, and 3 and for category 4 new medicines two or more tests on the major effects are required or well-documented material has to be submitted. Only tests on the major effects of the medicine on "new" cardinal symptoms are required for the new medicines in category 5. In general pharmacology studies, studies on nervous system, cardiovascular system, and respiratory system, should be covered, and technical requirements for toxicological studies were laid down in special monograph [27].

2.6 Europe

The European Union is composed of twenty-eight member states: Austria, Belgium, Bulgaria, Croatia, Cyprus, Czech Republic, Denmark, Estonia, Finland, France, Germany, Greece, Hungary, Italy, Ireland, Latvia, Lithuania, Luxembourg, Malta, the Netherlands, Poland, Portugal, Romania, Spain, Slovakia, Slovenia, Sweden, and the United Kingdom [29]. All member states are independently responsible for licensing herbal medicines; however, they can be added to the entire EU pharmaceutical market at once by following the regulations set by the European Medicines Agency (EMA) [30]. In contrast to the United States, a botanical drug is referred to as an herbal medicine. This is not to be confused with botanicals in the EU, which are plant derivatives used in food supplements that are instead regulated by the European Food Safety Authority (EFSA) [31]. As defined by Parliament under Directive 2001/83/EC [32],

Herbal Medicinal Product: any medicinal product, exclusively containing as active ingredients one or more herbal substances or one or more herbal preparations, or one or more such herbal substances in combination with one or more such herbal preparations.

Herbal substances: All mainly whole, fragmented or cut plants, plant parts, algae, fungi, lichen in an unprocessed, usually dried, form, but sometimes fresh. Certain exudates that have not been subjected to a specific treatment are also considered to be herbal substances. Herbal substances are precisely defined by the plant part used and the botanical name according to the binomial system (genus, species, variety and author).

Herbal preparations: Preparations obtained by subjecting herbal substances to treatments such as extraction, distillation, expression, fractionation, purification, concentration or fermentation. These include comminuted or powdered herbal substances, tinctures, extracts, essential oils, expressed juices and processed exudates.

An herbal medicine application can be authorized via one of four different authorization procedures: (1) centralized, (2) national, (3) mutual-recognition, and (4) decentralized. To attain market approval in all member states of the European Union, the centralized authorization procedure is followed. This procedure involves the submission of a single marketing-authorization application to the EMA. In this way, only one application is needed as opposed to twenty-eight separate applications. Drugs are required to be approved via the centralized authorization procedure if the herbal substance can treat HIV/AIDS, cancer, diabetes, neurodegenerative disease, auto-immune disease, or viral diseases [30]. Fulyzaq, as discussed in the introduction of this chapter, for example, would be required to follow the centralized authorization because it treats diarrhea in HIV/AIDS patients [5].

On the other hand, a sponsor could follow the national authorization procedure, which intuitively allows marketing in that single member state only, if approved. Sponsors may be obligated to use this procedure if the herbal medicine does not treat any of the diseases mentioned above and is therefore not qualified to follow the centralized procedure [30]. Under the national authorization procedure, sponsors must follow the regulations set by each specific member state in order to have an herbal medicine approved in each pharmaceutical market. Approval of clinical trials, for example, is a responsibility of each member state [33]. Table 2.2 indicates the authorities responsible for drug regulations in all twenty-eight countries of the EU [29]. There is also a mutual-recognition procedure, which allows the authorization granted in one member state to be used in other EU countries. Lastly, the decentralized procedure allows simultaneous authorization in more than one member state [30].

According to EMA, a medicinal product with any type of herbal substance can have access to the pharmaceutical market as one of three categories of applications. The first option is a *traditional use* herbal medicine: It does not require a prescription and has sufficient data regarding its safety and efficacy due to its extensive use in the EU. The regulation pathway for this category is allowed a simplified procedure to enter the pharmaceutical market. The sponsor must submit an application to its member state that includes data from pharmaceutical tests, details of all active herbal substance characteristics, any marketing approval or refusals obtained from a different country, and bibliographical evidence indicating the herbal medicine is safe and has been in use for at least 30 years prior. Second, an herbal medicine can be categorized under *well-established use*, which requires there to be scientific literature indicating that the active ingredient in the herbal medicine has had well-established use in the EU for at least 10 years, thus indicating established safety and efficacy. Herbal medicines in this category are usually granted marketing authorization by a member state or the EMA. Similar to traditional use herbal medicines, marketing authorization regulations for well-established use herbal medicines involve mostly bibliographic evidence [32]. Lastly, an herbal medicine can be *stand alone*, which requires the drug sponsor to provide its own data on the drug's safety and efficacy. The drug sponsor may also provide a combination of its own data with other cited

Table 2.2 Regulatory Authorities in the European Union Responsible for the Regulation of Human Medicines that are not Authorized by the European Medicines Agency

Country	Agency
Austria	Austrian Agency for Health and Food Safety
Belgium	Federal Agency for Medicines and Health Products
Bulgaria	Bulgarian Drug Agency
Croatia	Agency for Medicinal Products and Medical Devices of Croatia
Cyprus	Ministry of Health — Pharmaceutical Services
Czech Republic	State Institute for Drug Control
Denmark	Danish Health and Medicines Authority
Estonia	State Agency of Medicines
Finland	Finnish Medicines Agency
France	National Agency for the Safety of Medicine and Health Products
Germany	Federal Institute for Drugs and Medical Devices
Greece	National Organization for Medicines
Hungary	National Institute of Pharmacy and Nutrition
Ireland	Icelandic Medicines Agency
Italy	Italian Medicines Agency
Latvia	State Agency of Medicines
Lithuania	State Medicines Control Agency
Luxembourg	Ministry of Health
Malta	Medicines Authority
Netherlands	Medicines Evaluation Board
Poland	Office for Registration of Medicinal Products, Medical Devices and Biocidal Products
Portugal	National Authority of Medicines and Health Products
Romania	National Medicines Agency
Slovakia	State Institute for Drug Control
Slovenia	Agency for Medicinal Products and Medical Devices of the Republic of Slovenia
Spain	Spanish Agency for Medicines and Health Products
Sweden	Medical Products Agency
United Kingdom	Medicines and Healthcare Products Regulatory Agency

The EMA is only responsible for medicines authorized through the centralized authorization procedure.

sources. These herbal medicines are permitted to follow the centralized authorization procedure if they meet all requirements set by the EMA [34].

Whether the safety and efficacy data comes from bibliographic sources or the sponsor's own experiments, and whether the centralized or national authorization procedure is followed to access the pharmaceutical market, all herbal medicines must provide the same information as outlined in Parliament Directive 2001/83/EC: the binomial scientific name of the herbal substances and/or herbal preparations must be provided along with the molecular structure and formula; the process and location from which the plant is obtained as well as a summarized explanation of how it is manufactured (e.g., solvents used, purification stages, and so on); any experimental data obtained when testing the herbal substance(s) and/or herbal preparation(s) [32].

2.7 Japan

Pharmaceuticals and Medical Devices Agency (PMDA) is the name of the authority responsible for creating regulations for all drugs in the Japanese market. In Japan, there is not a particular route taken specifically for the addition of botanical drugs to the pharmaceutical market. Instead, as with all other drugs in Japan, sponsors must follow the same procedures briefly summarized in Figure 2.2 [35].

When experimental trials begin, the PMDA assesses the safety and efficacy data to ensure proper ethical and scientific standards are met. A group of qualified pharmaceutical and health professionals evaluate the drug's quality, pharmacology, pharmacokinetics, toxicology, clinical implications, and biostatistics. Each member of the review team offers their opinions about their area of expertise, and they also exchanges opinions with other experts in the field. This ensures a more effective review for each drug being evaluated. In 2013, a total of 7,118 drugs were approved: 4,008 were prescription drug, 916 were OTC drugs / BTC drugs, 166 were in vitro diagnostics, and 2,028 were quasi-drugs. The specific number of botanical drugs is not noted. Again, as with the previous pharmaceutical markets discussed, Japan has quality-control procedures to make sure all batches of drugs are equivalent and the drug manufacturing sites are appropriate. Interestingly, when it comes time to submit a new application, Japan only accepts those written in Japanese, however they do accept data attained from clinical trials practiced outside of Japan. In fact, in 2007, Japan approved twenty-four medical device applications that used Japanese clinical data, and twenty medical device applications that used foreign clinical data only. In 2011, Japan approved fourteen medical device applications that used Japanese clinical data only, and

Figure 2.2 *This flowchart illustrates an outline of the procedures for drug authorization in Japan. In Japan there is not a specific set of regulations for the approval of marketing for botanical drugs.*

thirty-eight medical device applications that used foreign clinical data only, a more than twofold increase in approvals. Only *after* a new drug has entered the Japanese market does PMDA use relief services for adverse effects caused by the new drug [35].

2.8 Brazil

The use of botanical drugs is widely accepted in Brazil due to their extensive biodiversity, which makes up 22% of total species in the world. The governing agency responsible for regulation of products that can interfere with health, such as food, drugs, cosmetics, airports, and borders, is Agência Nacional de Vigilância Sanitária, or ANVISA. The National Policy of Integrative and Complementary Practices in the Public Health System (PNPIC) through Ordinance 971/2006, and the National Policy on Medicinal Plants and Herbal Medicines (PNPMF) through Decree 5813/2006 state regulations for herbal medicines [36].

Recently, ANVISA updated their regulations. They reviewed documents from the World Health Organization to survey international regulations of regions like the EU, Australia, and Canada. Then, they took the most important points from each, and developed an improved version of their own. Now, ANVISA regulations separate herbal medicines into two categories: herbal medicines, which substantiate their safety and efficacy through clinical trials, and traditional herbal medicines, which verify safety and efficacy through proof of long-term usage. Traditional herbal medicines can only be used topically or orally. These updates were clearly inspired by the EU's regulations.

To market an herbal medicine, manufacturers must gain approval by the procedures outlined in RDC 26/2014. A dossier must be submitted reporting the drug's production, quality control, safety and efficacy, clinical data, and labeling. Safety and efficacy of herbal medicines must be proven with clinical data, while that of traditional herbal medicines can undergo a simplified registration procedure. Prior to the commencement of Phase I, II, and III of clinical trials for herbal medicines, manufacturers must attain permission from ANVISA and an ethics committee to ensure effective and ethically sound procedures are being used, respectively. To qualify for simplified registration procedures, the medicinal plant within the drug must be listed on an official Brazilian list, which currently holds twenty-seven species that qualify for simplified procedures, and describes the dossier requirements for each. Drugs categorized under traditional herbal medicine have the option to be authorized by the simplified procedures or bibliographically proving longevity.

Along with safety and efficacy information, manufacturers are required to summarize procedures by which contamination is tested. The results of these tests must be included to validate the purity of every batch. Additionally, under RDC 26/2014, registration of herbal medicine is valid for only five

Table 2.3 Make-up of Brazil's Herbal Medicine Market

Herbal Medicine Composition	Number of Herbal Medicines
Single medical plant	357
More than 1 medical plant	25
Total	382

As of 2014, there are a total of 382 medicines in Brazil; 357 of them are made up of only one species of medicinal plant, and the remaining 25 are made up of more than one medicinal plant. Only 25% of the total species originated from South America.

years, so manufacturers must re-apply for a five-year renewal by proving that its safety, efficacy, and quality remains unchanged. Table 2.3 shows the number of herbal medicines that have been successfully registered in the Brazilian market [36].

2.9 Regulatory challenges for developing botanical drugs

The most common challenge for manufacturing companies that develop herbal drugs is adhering to the regulations set by government agencies to confirm the purity of the herbal substance and to keep the potency equivalent in all batches of the same drug. The challenge comes from trying to convert a traditional remedy, with known function due to continuous use over many generations, into a modern medicine. Often times the scientific evidence needed to standardize a medicinal herb is scarce, and since each herb has unique, unknown active ingredients, the data is difficult to gather. However, gathering this data is crucial to meet the industrial norm set by synthetic drugs that allows healthcare professionals and patients to trust their prescriptions. It is also needed for botanical drugs to be considered a legitimate form a treatment.

There are many variables to consider when attempting standardizing a medicinal herb. First, the potency of the raw herb can vary depending on where it is grown and harvested. Even if grown in the same exact location, the potency can change from time to time due to climate change or genetic changes in the species. The process by which the raw material is stored can also affect herb potency. Sponsors must utilize appropriate laboratory tests to verify that there is a constant level of potency in each batch of drug produced. Another challenge to be considered when processing medicinal herbs is its purity. There is a possibility of chemical and microbial contaminants; they must be removed to develop the drug without altering the herb's potency and function. For proper standardization, herbs must be grown using good agricultural practice (GAP). But even when being most careful, there is still a chance of variability in herbal substances. There is no methodology that promises constant herbal content.

After the drug is produced, its safety must be substantiated, most commonly through clinical trials. This leads to the second challenge drug manufacturers face, which is creating the ideal trial that is randomized, double-blind, and placebo-controlled. The difficulty comes in creating a drug that is odorless and tasteless, so the trial patients are not aware of what medicinal compound they are being given. Many botanical substances have very distinct smells and tastes, so it becomes a challenge to remove those properties without altering or hindering the product's function, or the quantity of the medicinal compound in each dosage. Some of the most common issues in clinical trials for botanical drugs are inexplicit enrollment criteria, deficient sample sizes with high dropout rates, and poor follow-ups [37].

References

1. M. Ekor, The growing use of herbal medicines: Issues relating to adverse reactions and challenges in monitoring safety, *Frontiers in Pharmacology 4* (2014).
2. J.B. Calixto, Efficacy, safety, quality control, marketing and regulatory guidelines for herbal medicines (phytotherapeutic agents), *Brazilian Journal of Medical and Biological Research 33* (2000), pp. 179–189.
3. F.X. Liu and J. Salmon, Comparison of herbal medicines regulation between China, Germany, and the United States. *Integrative Medicine: A Clinician's Journal, Integrative Medicine: A Clinician's Journal 9* (2010), pp. 42–49.
4. Center for Drug Evaluation and Research. Veregen Ointment Printed Labeling. (NDA 021-902). Available at https://www.accessdata.fda.gov/drugsatfda_docs/nda/2006/021902s000_prntlbl.pdf.
5. Fulyzaq delayed-release tablets Printed Labeling. Reference ID: 3238051. Available at https://www.accessdata.fda.gov/drugsatfda_docs/label/2012/202292s000lbl.pdf.
6. US-FDA (CDER), Botanical Drug Development Guidance for Industry, (2016), p. 34.
7. Consolidated federal laws of Canada, Natural health products regulations. Available at http://laws-lois.justice.gc.ca/eng/regulations/SOR-2003-196/.
8. Central Drugs Standard Control Organization, Gazzete of India GSR 702 (E), REGD. NO. D. L.-33004/99 (2013).
9. Central Drugs Standard Control Organization, Gazzete of India No. 741, REGD. NO. D. L.-33004/99 (2015).
10. Ministry of Health and Family Welfare, Drug and Cosmetics Act, 2005.
11. Drugs & Magic Remedies Objectionable Advertisement Act, 21 (1954), p. 10.
12. The Biological Diversity Bill, 2000. Available at http://envfor.nic.in/legis/others/biobill.html.
13. The Indian Forest Act, 1927. Available at http://envfor.nic.in/legis/forest/forest4.html.
14. Indian Patent Act 1970-Sections. Available at http://ipindia.nic.in/writereaddata/Portal/ev/sections-index.html.
15. D.A. Narayana and C. Katiyar, Draft amendment to drugs and cosmetics rules to license science based botanicals, phytopharmaceuticals as drugs in India, *Journal of Ayurveda and Integrative Medicine 4* (2013), pp. 245–246.
16. Complementary Medicines Australia. 2017/18 Federal pre-budget submission. Complementary Medicines Australia, Mawson, Australia, 2017.
17. Australian regulatory guidelines for complementary medicines (ARGCM). Department of Health, Therapeutic Goods Administration, 2018.
18. Health, Therapeutic Goods Act 1989, 21 (1989).
19. Health, Therapeutic Goods Regulations 1990, Statutory Rules No. 394 (2013).
20. How the TGA regulates. Available at https://www.tga.gov.au/how-tga-regulates.
21. Medicines and TGA classifications. Available at https://www.tga.gov.au/medicines-and-tga-classifications.
22. An overview of the regulation of complementary medicines in Australia. Available at https://www.tga.gov.au/overview-regulation-complementary-medicines-australia.

23. L. Sansom, W. Delaat and J. Horvath, Review of Medicines and Medical Devices Regulation: Stage Two. Report on the regulatory frameworks for complementary medicines and advertising of therapeutic goods, Australian Government, Department of Health, Canberra, Australia, 2015.

24. Regulation of Complementary Medicines in Australia | Issues Magazine. Available at http://www.issuesmagazine.com.au/article/issue-september-2008/regulation-complementary-medicines-australia.html.

25. WHO Western Pacific Regional Office (WPRO), Communication with WHO Geneva, Switzerland, 1996.

26. X. Wang, Traditional Herbal Medicines around the Globe: Modern Perspectives. China: Philosophical Basis and Combining Old and New. Proceedings of the 10th General Assembly of WFPMM, *Swiss Pharma 13*(11a) (1991), pp. 68–72.

27. J.-Y. Tang, L. Ma, L. Zhang, J. He and H. Zhang, Chinese traditional medicines face crucial challenges in the new regulatory science: Approval and reflection on Chinese Herbal Medicines, *SJAMS 1* (2014), pp. 373–378.

28. People's Republic of China, Drug Administration Law of the People's Republic of China, 1984, p. 21.

29. European Medicines Agency: EU Member States—National competent authorities (human). Available at http://www.ema.europa.eu/ema/index.jsp?curl=pages/medicines/general/general_content_000155.jsp&mid=WC0b01ac0580036d63.

30. European Medicines Agency: About Us—What we do. Available at http://www.ema.europa.eu/ema/index.jsp?curl=pages/about_us/general/general_content_000091.jsp&mid=WC0b01ac0580028a42.

31. Botanicals. Available at http://www.efsa.europa.eu/en/topics/topic/botanicals.

32. Directive 2001/83/Ec of the European Parliament and of the council of 6 November 2001 on the community code relating to medicinal products for human use, Official Journal L-311, Europe, 2004.

33. EMA, The European Regulatory System for Medicines and the EMA. A consistent approach to medicines regulation across the European Union, 437313, European Medicines Agency, 2014.

34. European Medicines Agency: Human regulatory—Herbal medicinal products. Available at http://www.ema.europa.eu/ema/index.jsp?curl=pages/regulation/general/general_content_000208.jsp.

35. Reviews | Pharmaceuticals and medical devices agency. Available at http://www.pmda.go.jp/english/review-services/reviews/0001.html.

36. A.C. Bezerra Carvalho, L.S. Ramalho, R.F. de Oliveira Marques and J.P. Silvério Perfeito, Regulation of herbal medicines in Brazil, *Journal of Ethnopharmacology 158* (2014), pp. 503–506.

37. Y. Liu and M.-W. Wang, Botanical drugs: Challenges and opportunities, *Life Sciences 82* (2008), pp. 445–449.

3

Bioprospecting, Botany, Biodiversity, and Their Impact on Botanical Drug Development

Suma Krishnaswamy

Contents

3.1 Introduction to bioprospecting

Bioprospecting has been defined as the process of development of traditional medicines into commercial products. While this definition sounds very simple, the route to developing viable products from nature is intensively tortuous. The process of foraging has evolved parallel to the evolution of man. As the primary provider in the food chain, all forms of primitive or advanced plants have been targets for various forms of prospecting: as food for nourishment, as essences that induce well-being, as the plant spirit for therapy, as extracts for healing. The goal of bioprospecting has been to identify prospective botanicals/molecules with the highest level of pharmacological activity and the least level of toxicology that could serve as (a) new drug for a new malady, (b) new drug for a chronic disease, (c) new drug as a substitute/alternative to an existing remedy that has run scarce, (d) existing drug re-aligned to treat a different malady. The subjects of intense investigation toward this end have been all classes of plants, animals, microbial organisms, minerals, saprophytic or parasitic forms of plants and animals. Several leads have been generated and taken to commercialization as molecules. Prominent botanical drugs that have matured to commercial drugs over the past 30 years are from *Artemisia annua*, quinghaosu, (artemisinin) for malaria, *Salacia reticulata* (Salaretin, salacinols) for diabetes, *Rauwolfia serpentina* (reserpine) for hypertension, *Psoralea corylifolia* (Psoralens) for vitiligo, *Holarrhena antidysenterica* alkaloids in amoebiasis, *Commiphora guggul* (guggulsterones) as hypolipidemic agents, *Mucuna pruriens* (L-Dopa) for Parkinson's disease, *Piper nigrum* (piperidines) as bioavailability enhancers, *Bacopa monnierii* (baccosides) in mental retention, *Picrorrhiza kurroa* (picrosides) in hepatic protection, *Phyllanthus amarus* (hypophyllanthins) as antivirals, *Curcuma longa* (curcumine) in inflammation, *Withania somnifera* (withanolides) as immunomodulators. These drugs have been mentioned quite elaborately in Sushruta Samhita, which also catalogues about 1,308 entries relating to about 700 plant species and is believed to have been written around 1000 BC. Even plants like *Orthosiphon pallidus* (Sanskrit: Arjaka) that are restricted to the Arabian peninsula, Pakistan, India have been described. It was only in 2005 that *Orthosiphonus stamineus* acquired importance as java tea, a traditional herbal medicinal adjuvant used to increase the amount of urine and achieve flushing of the urinary tract in minor urinary tract complaints and a monograph was submitted by the HMPC (Committee for Herbal Medicinal Products), European Medicines Agency. A well-documented value-based example of exploratory research isolation of an enzyme named Taq polymerase from the microscopic denizens of the hot springs of Yellowstone National Park called *Thermus aquaticus* during the late 1980s [1]. A polymerase is an enzyme that catalyses the formation and repair of DNA and RNA from an existing strand of DNA (or RNA) serving as a template and has become an indispensable component of any molecular biology experiments today.

Even though vast diversity in living and nonliving forms is available as a resource, bioprospecting has returned few new hits/leads that could quickly be evaluated for their therapeutic properties and taken to the level of commercialization. It is worth a deep consideration as to why there are so few successes in recent times despite high technological advancements and sophisticated techniques. The traditional remedies advocated tens of thousands of years ago are still effective today despite the physical, emotional, environmental shifts, and high stress levels that today's man is exposed to. What did the ancient sages and healers of the Indian subcontinent, the Oriental and Western Herbalists, Shamans, Red Indians, and other cultures learn from nature, their own intuition, and observation and apply successfully in their formulations that we are missing?

Biodiversity of various countries (comprising not only of various orders of the plant kingdom but also organisms such as leeches, marine forms, parasitic and saprophytic fungi) has been a central source of highly effective indigenous medicines. The practice of harnessing nature for the benefit of man and his domesticated animals has been prevalent from Indo-Vedic times. The behavior of suffering animals that instinctively head toward the right herb to find an immediate relief for its malady has probably been the way early healers discovered the medicinal value of different medicinal herbs (Zoopharmacognosy). Charaka Samhita (Circa 400–210 BC) postulates several therapies striving to restore the balances between the mind, body and soul or Intellect using natural resources such as plants, minerals and organisms [2]. These intuitive therapies accompanied by specific chanting of Mantras with particular intonation were successfully used as very powerful healing methods more than 10,000 years ago. Unlike modern medicine, elements in nature are integrally incorporated into their approaches. The Ayurvedic form of healing enriches the qualities of life of a person through the different sensations of intuitive feeling, sound, touch, vision, taste, and smell. Elements of nature and Cosmic Intellect are an integral part of healing in most traditional forms of healing in cultures of the Appalachian mountain tribes, Native Americans, Mayas, Aztecs, and others. Complementary and Alternative Medicine and all traditional healing systems consistently believe in the strong connect between the Science of Medicine and the Spiritual aspect which facilitates and hastens healing. For example, we all possess innate healing capacities; healing occurs from within in the context of healthy lifestyles and ecosystems; we can learn from nature by careful visual, olfactory or auditory observation of the plants themselves (morphologic and physical features, energetics, habitat associations).

Science and Spirituality, being two sides of the same coin, reflect the ambivalence and duality that prevails in the Universe. While Science poses questions and insists on the tangibles, Spirituality provides the esoteric answers. This Universal Duality is explained by ancient sages in Ayurveda as Prakriti and Purusha, Oriental concept of Yin and Yang that is, Nature and Essence. Matter takes a physical form (Purusha) that is propelled by its power of manifestation (Nature or Prakriti). The physical form provides valuable hints about the Prakriti or nature of the physical form either as similarities or contrasts which can be interpreted in various ways: Plant morphology, the target organ based on appearance, phytochemical moieties based on specific colours, unique markings on leaves, flowers or fruits, shape or structural architecture adopted that suggest their potential activity.

This principle is true even in the microscopic plant cell where each individual cellular unit has a physical presence and an inbuilt intellect that modern Botany calls Totipotency. This inbuilt intellect leads the cell into acquiring a form based on the set of stimuli it is presented with. For instance, growth regulators as a source of stimulus, play a major role in orchestrating cell responses in tissue culture. A plant being in the material form and its unique combination of nutritive juices contains the Essence or the expression of its intellect in the form of potency or *Virya*. The Cosmic Intellect of the plants allows them to perceive external stimuli and elicit complex and inexplicable morphological and phytochemical responses. While this *Virya* or Potency of therapeutic plant juices is a boon to humanity, it also has been the bane of plant kingdom, the very reason for its vulnerability to exploitation and eventual extinction. It is imperative for mankind to practice responsible and sustainable Bioprospecting to ensure that quest for the Golden Egg does not kill the Goose. Botany, Bioprospecting, and Biodiversity Conservation constitute the holy triad that form the core of sustainable use of herbs.

3.2 Techniques for effective bioprospecting

The relentless search for prospective botanicals either from new plant species or for new intermediates from known sources or new therapeutic application of known compounds has been a continuous process and has taken several diverse routes. The common approaches adopted by different investigators can be listed as follows:

- Zoopharmacognosy
- Ethno pharmacology/indigenous knowledge
- Doctrine of Signatures
- Ecological interactions and adaptations
- Shot–gun or random approach

3.2.1 Zoopharmacognosy

Zoopharmacognosy is an emerging applied science, defined as the formal study of plants used by animals in their native habitats which are believed to use to "treat" their own illnesses. (McGraw-Hill Concise Dictionary of Modern Medicine ©2000). Cattle, primates, monarch butterflies, and sheep apart from knowing how to ingest the right medicinal plants to treat their illness, also knew which plant to avoid. Goats and sheep are known to steer clear of unpalatable and fetid smelling *Adathoda vasica*, which contains Vascicine, an alkaloid toxic to termites, mosquitoes, and cold blooded animals including fish. Gary Le Mon, while listing out different instances of animal-adopted herbs for self-medication observed that animals do not seek out medicinal herbs until they are afflicted with a disease condition that needs curing [3]. Traditionally, cultures across the globe have learned by keen observation of behavior of wild animals to decipher their illness and the herbs or cures that they instinctively seek out. Biologists witnessing animals eating foods not part of their usual diet, realized the animals were self-medicating with natural remedies.

Esther Inglis-Arkell [4] wrote about researcher Holly T. Dublin who observed a pregnant African elephant for over a year to make an astounding discovery. The elephant kept regular dietary habits throughout her long pregnancy term of 22 months but the routine changed abruptly toward the end of her term. Heavily pregnant, the elephant set off in search of a shrub that grew 17 miles from her usual food source. The elephant chewed and ate the leaves and bark of the bush, then gave birth a few days later. The elephant, it seemed, had sought out this plant specifically to induce her labour. The same plant (a member of the borage family) also happens to be brewed by Kenyan women to make a labor-inducing tea.

In another incident concerning Tanzanian chimpanzees, it was observed that chimps ailing with stomach ache chose leaves from *Aspilia africana*, a semi woody Compositae member with a somewhat aromatic carroty smell and bright lemon yellow-colored flowers. The entire plant is covered with numerous stiff trichomes. The chimps carefully fold up the leaves, roll them around their mouths before swallowing whole. The prickly leaves scour parasitical worms from the chimp's intestinal lining. Further these same animals were observed to chew the intensely bitter pith of Vernonia with distinct expression of distaste. Healthy animals were not seen seeking out these two plants, both of which have been used in Tanzanian Ethnomedicine for diarrhea and intestinal parasites [5]. Télesphore Benoît Nguelefack et al. [6] reported that *Aspilia africana* (Asteraceae) is used in Cameroon ethnomedicine for the treatment of stomach ailments. The methanolic leaf extract of *A. africana* was investigated against gastric ulcerations induced by HCl/ethanol and pylorus-ligation in Wistar Rats. With both methods, the extract inhibited gastric ulcerations in a dose-related manner. Oral administration of the plant extract at the doses of 0.5 and 1 g/kg reduced gastric lesions induced by HCl/ethanol by 79% and 97% respectively. The extract at the dose of 1 g/kg reduced gastric lesion in the pylorus ligated rats by 52% although the gastric acidity remained higher as compared to the control. These findings show that methanol extract of the leaves of A. africana possess potent antiulcer properties.

Similarly, Farombi and Owoeye reported that the leaves of *Vernonia amygdalina* (bitter leaf) are used in Cameroon as a green vegetable or as a spice in soup, especially in the popular bitter-leaf soup (ndolé) [6]. It is used for calming down toothache by direct application on a cavity. Saponins and alkaloids, terpenes, steroids, coumarins, flavonoids, phenolic acids, lignans, xanthones, anthraquinones, edotides, and sesquiterpenes have been extracted and isolated. Extracts of the plant have been used in various folk medicines as remedies against helminthic, protozoal, and bacterial infections with scientific support for these claims.

It is not just the primates or mammals that are endowed with high discerning capacity with respect to ferreting out the perfect plant for its therapy. Some species of South American parrot and macaw are known to eat soil with high kaolin content. The parrots eat fruit seeds such as apple which contains cyanide and other toxins. The kaolin clay absorbs and flushes out the toxins. Kaolinite clay occurs in abundance in soils that have formed from the chemical weathering of rocks in hot, moist climates—for example in tropical rainforest areas Kaolin has been used for centuries in many cultures as a remedy for human gastrointestinal upset. Kaolin is also eaten for health or to suppress hunger, in a practice called Geophagy.

So what Providence gives these mute animals such wonderful knowledge that they even know the best practices to find, harvest and ingest the drug in a most efficient drug delivery form? While they certainly do not seem to be adopting the shot-gun approach, it is not clear which method they employ to arrive at the correct remedy. It is possible that the flower color implying presence of flavonoids, subtle aromas indicating terpenes and volatiles, tactile clues like texture of the plant parts or the taste of the natural resource leads the animal or bird to its cure. That implies that the systematic analysis or elimination of a plant or a natural resource as a potential cure is unconsciously conducted by the animals' at a sensory level. Is it so simple or is it some inner craving for some deficient molecules that attract the animal to the therapeutic or curative source?

3.2.2 Ethno pharmacology/indigenous knowledge

Successful civilizations have tried to evolve therapeutic and curative practices best suited for survival under harsh weather conditions, threat from natural aggressors, diseases and pests. Many of the systems so developed have been handed down through generations by word of mouth or some kind of crude or sophisticated documentation.

While there are numerous traditional knowledge banks such as Unani, Siddha, Traditional Chinese Medicine, the methods and practices of native drug development followed by two major schools of thought—Ayurveda and Herbalism—have been considered here.

3.2.2.1 Ayurveda

Ayurvedic experts suggest a reverse-pharmacology approach focusing on the potential targets for which ayurvedic herbs and herbal products could bring tremendous leads to ayurvedic drug discovery. Although several novel leads and drug molecules have already been discovered from ayurvedic medicinal herbs, further scientific explorations in this arena accompanied by customization of present technologies to align with ayurvedic drug manufacturing principles would greatly facilitate a standardized ayurvedic drug discovery.

Ayurveda describes a Plant being, as being composed of seven Dhatus or Seven Planes that can be correlated to a similar composition of a Human being. The seven Dhatus are: Plant sap: *Plasma* | Resin: *Blood*| Gum: *Fat* | Softwood: *Muscle*| Bark: *Bones*| Leaves: *Nerves*| Flowers: *Reproductive organs*.

Each of these Dhatus or plant organs possesses therapeutic potential toward the corresponding human plane or Dhatus or tissue. This very basic postulate provides valuable clues to investigators trying to assign the relevant therapeutic plant tissue to a particular human organ or disease. Further Ayurveda applies the TriDosha concept to plants as well, classifying them as Kapha plants (characterized by luxuriant growth, abundant leaves and sap, dense, heavy, succulent leaves, and contain lot of water), Vata plants (possess sparse leaves, rough cracked bark, gnarled branches, and spindly growth with very little sap coursing through its body) and Pitta plants (characteriszed by brightly colored plants with spectacular flowers, and moderate in strength and sap), often poisonous or creating a burning feeling. According to Vagbhata, one of the three classic writers of Ayurveda, the karma *vidhana,* or primary mechanism of action, could come from the plant's "taste, potency, post-digestive effect, qualities or unique property. Whichever trait is most powerful in the plant lessens the other traits and becomes the cause of action" -*Astanga Hrdayam*, IX.22-23 [7].

3.2.3 Herbalism: doctrine of signatures

The Greek physician Galen (AD 129–200) devised the first pharmacopoeia describing the appearance, properties, and use of many plants of his time. Paracelsus aka Theophrastus Bombastus von Hohenheim (1493–1541), a Swiss physician, alchemist, philosopher, and father of modern chemistry, observed that the qualities of plants are often reflected in their appearance. He theorized that the inner nature of plants may be discovered by their outer forms or Signatum/signatures. He applied this principle to food as well as medicine, remarking that "it is not in the quantity of food but in its quality that resides the Spirit of Life." It is referred to as the Doctrine of Signatures, and in a time when knowledge of medicinal plants was passed on by word of mouth, it proved a practical way of remembering a plant's properties. This practice was in part a spiritual one as it was believed that God marked his creations with a clear indication or "signature" of its purpose. Groups of plants sharing the same signature were thought to have similar healing properties or have a healing effect on similar parts of the body [8,9].

Francis Bacon approached this theory with a reductionist science, in which the whole is reduced to the parts with rational and scientific study. However, he also taught that it was necessary to think holistically, to put the pieces back together. The way to do this, he taught, was to think by analogy, for causal, rational similarities could trace out relationships in nature. For instance, he said, there must be a developmental relationship between the womb and the scrotum, due to similar shape. Actually, this method is unconsciously used in science. Darwin, for instance, reasoned from the morphological similarities in birds on the Galapagos to arrive at the Theory of Evolution.

Goethe's (1749–1832) in his first major scientific work, the *Metamorphosis of Plants*, tried to establish a science based on analogical thought. He was the first to observe that the flower structures were modified leaves, noting "From first to the last, a 'Plant' is nothing but 'leaf' which traverses the dynamic path of cotyledons, stem, leaves, petals, stamens, pistil, fruit and seed which is called Metamorphosis of Plants" *and the cranium was modified vertebra.* Goethe wrote in *Story of My Botanical Studies* (1831): "The ever-changing display of plant forms, which I have followed for so many years, awakens increasingly within me the notion: The plant forms which surround us were not all created at some given point in time and then locked into the given form, they have been given... a felicitous mobility and plasticity that allows them to grow and adapt themselves to many different conditions in many different places." This demonstrated the thought process that the plant form adapts to environment, that the same plant species may express itself differently in different situations. He further wrote that the plants progressed from crude primitive forms to advanced forms of varying complexity, striving for near-perfection in a process called Intensification and Polarity, which is evident as alternating forces of expansion and contraction, tracing the development of seed expanding to germinate into a stem-leaf or plant, contracting stem to sepal, expanding sepal to petal, contraction of petal to pistil and stamen, expansion from reproductive organs to fruit and finally contraction of fruit to seed. He also observed that the characteristic expression of underlying idea in any plant is a coordinated result of the "law of inner nature" which determines the constitution of the plant and the Law of Environment whereby the plant has been modified. Rudolf Steiner the Father of Anthroposophy has attempted to perpetuate Goethe's approach.

Plants have been used fairly accurately for their therapeutic worth for thousands of years by ancient ayurvedic sages and herbalists who relied on a few parameters to fit a plant to the malady. A plant, in its physical manifestation, very cleverly hides its true attributes in often unrecognized indecipherable signatures and herbalists succeeded in identifying therapeutic plants based on valuable clues that lay camouflaged in the plant parts. Plant parts starkly similar to human body parts possessed bioactivity with respect to that organ. Prominent examples of such phyto-therapies where "the plant part expresses its organ of therapy" are walnut, kidney beans, tomato, carrot, where the plant parts mimicked specific human organs (Figure 3.1). Plants may lack several of the evolutionary advantages enjoyed by animals, but they are enigmatic and mysterious in their ways of functioning, expression and self-preservation.

Paracelsus also believed that plants grow where they are most needed; for example, dock leaves were used to treat the sting from a nettle plant, and these two plants were often found growing close together. Aspects of the plant which were thought to give indications to its use include:

- Habitat of the plant
- Colour—of flower, fruit, root, or stem
- Shape
- Texture
- Odour

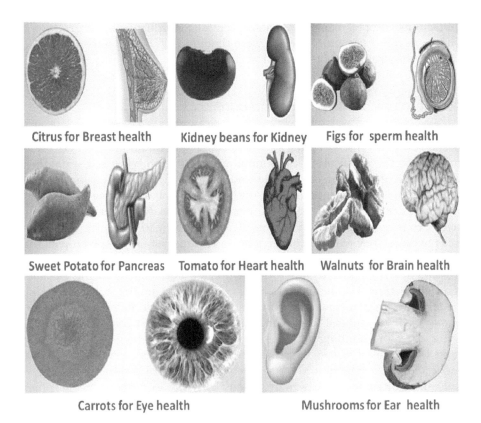

Citrus for Breast health   Kidney beans for Kidney   Figs for sperm health

Sweet Potato for Pancreas   Tomato for Heart health   Walnuts for Brain health

Carrots for Eye health   Mushrooms for Ear health

Figure 3.1 *Fruits and vegetables mimic human organs and also possess therapeutic activity toward that organ. (Adapted from The ancient "Doctrine of Signatures" suppressed by the establishment, Available at https://www.richardcassaro.com/the-ancient-doctrine-of-signatures-suppressed-by-the-elite.)*

Some of the most reputed examples of the doctrine of signatures from that time include lungwort, whose spotted leaves were believed to resemble a diseased lung; walnuts, which were considered to be shaped like the human brain; and ginseng root, which was used to assist male sexual vitality due to its resemblance to male reproductive anatomy.

Modern herbalism has confirmed some of these earlier observations; for example, lungwort is an expectorant herb used to help clear mucus from the lungs, and walnuts with their omega 3 fatty acids are considered beneficial for brain health.

The Sioux people formerly inhabiting northern Nebraska and southern South Dakota were aware that thyme-leaf spurge (*Euphorbia serpyllifolia*), a small, sprawling annual of sandy soils, "was boiled and the decoction drunk by young mothers whose flow of milk was scanty or lacking, in order to remedy that condition." This may be an example of the "belief in signs," where characteristics of a plant reveal its potential medical use—spurges have a white, latex-like sap and therefore were thought suitable for nursing mothers in need of increased milk production [10].

3.2.4 Ecological interactions and adaptations

Michael Tierra correlated pharmacological significances of herbs growing under seven diverse habitats and geoecological conditions [11]. (1) Herbs growing on the north side of a mountain are more tonifying and strengthening because of the increased power and stamina they must develop in order to survive in a more difficult growing area. (2) Herbs growing in lowlands or near water are more beneficial for urinary diseases. (3) Herbs growing in high, dry desert regions tend to be of more benefit to the spleen and pancreas because they would help the traditional function of the spleen-pancreas according to Chinese medicine, to transform moisture in the body. (4) Herbs growing in fertile, nitrogenous soil would be of more benefit for the digestion and assimilation. Also herbs whose job is to fix nitrogen in the soil, including the leguminous plants such as clover and alfalfa, would help in our metabolism of proteins and cell formation. (5) Herbs growing in cold, harsh weather tend to be more building and heating in contrast to herbs growing in hotter, more temperate climates, which would tend to be more eliminating. While both can be detoxifying in their respective ways, there is a subtle but important tendency manifested by herbs found growing in different geographical localities. (6) The tastes of herbs are important indicators of their properties. The sweet taste is nutritive; pungent is dispersing; salty taste influences water balance and digestion; sour is digestive and cooling; bitter is cooling and detoxifying. Many herbs have a number of tastes and therefore a number of properties. (7) Aromatic herbs influence the subconscious mind and its multitudinal psychological processes.

Strobel G A and Bryn Daisy enumerated four main criteria for selection of medicinal plants in their review of bioactive rich plant-endophytic associations [12]. They are: (1) Plants from unique environmental settings, especially those with an unusual biology, and possessing novel strategies for survival are seriously considered for study. (2) Plants that have an ethnobotanical history (use by indigenous peoples) that are related to the specific uses or applications of interest are selected for study. These plants are chosen either by direct contact with local peoples or via local literature. Ultimately, it may be learned that the healing powers of the botanical source, in fact, may have nothing to do with the natural products of the plant, but of the endophytes (inhabiting the plant). (3) Plants that are endemic, that have an unusual longevity, or that have occupied a certain ancient land mass, such as Gondwanaland, are also more likely to lodge endophytes with active natural products. Srobel et al. cite the example of an aquatic plant, *Rhyncholacis penicillata*, collected from a river system in southwest Venezuela where the harsh aquatic environment subjected the plant to constant beating by virtue of rushing waters, debris, and tumbling rocks and pebbles [13]. Though the force of water caused injury to the plant, the plant population appeared to be healthy, possibly due to protective endophytic synergism. This environmental biological clue indicated the unique situation where a higher organism is protected by a microbial life form in exchange for perhaps nutrition. A potent antifungal strain of *Serratia marcescens* isolated from *R. penicillata* was shown to produce oocydin A, a novel antioomycetous compound having the properties of a chlorinated macrocyclic lactone. Oocydin A may be used as a biocontrol agent to combat Oomycetes such as *Pythium* and *Phytophthora*.

The ecological approach to select plant material is based on the observation of interactions between organisms and their environment that might lead to the production of bioactive natural compounds. Plants that attract bees and butterflies possess high amounts of distinct volatiles, which may be helpful in imparting a sense of wellbeing in humans. The hypothesis underlying this approach is that secondary metabolites, e.g., in plant species, possess ecological functions that may have also therapeutic potential for humans. For example, metabolites involved in plant defence against microbial pathogens may be useful as antimicrobials in humans, or secondary products defending a plant against herbivores through neurotoxic activity could have beneficial effects in humans due to a putative central nervous system activity.

3.2.5 Shot-gun or random approach

As the term signifies, this method is a rather indiscriminate, unplanned, and unorthodox approach that holds more promises than guarantees toward lead generation. Cragg et al., while sharing the mammoth herbal screening experience at the National Cancer Institute (NCI), used the term "shot-gun approach" to describe the huge numbers of random samples screened over a 20 year period [14]. The quest for anti-cancer compounds spanned about 114,000 extracts from around 35,000 plants, mainly collected in temperate regions. But quite interestingly, none of the only three clinically active anti-cancer drugs screened as part of this exhaustive exercise originated on the basis of long-standing traditional use in spite of *Taxus* spp possessing long ethnobotanical history. What aspect of the quest was missed in the experimental design that threw out such a minute number of drug leads and drastically skewed results? Is it possible that the idea of focussing only on temperate plants with their environment-induced, naturally subdued rate of metabolism was incorrect? Bills et al. observed a metabolic distinction between tropical and temperate endophytic associations: Tropical endophytes not only yielded natural compounds with higher activity but also larger number of active secondary metabolites than did the temperate endophytes [15].

It is reported that NCI realigned its studies to encompass tropical and subtropical species with ethnobotanical identity, taxonomic diversity, and global distribution. Between 1986 and 1993, 21,881 extracts were extracted from over 10,500 natural samples for their potential activity against human immunodeficiency virus

(HIV). About 90% of the medicinal-plant derived extracts displaying significant activity against HIV were aqueous extracts. It was inferred that the anti-HIV activity was due to the tannins and polysaccharides that are majorly part of the aqueous extractives. This method of Bioprospecting is not only intensive in terms of expense, effort and time; it fails to guarantee commensurate results.

3.3 Botany

Botanical investigations form an integral part of exploring, identifying, understanding and extrapolating critical information about Pharmacology of plant resources or Phyto-Capital. Laurence M.V. Totelin citing ancient Pharmaco-Botanists like Theophrastus's *Enquiry into Plants* (fourth century BCE), Dioscorides's Materia *Medica* (first century CE Hippocratic Corpus (fifth and fourth centuries BCE) observes that pharmacology and botany today are two distinct scientific fields, but in antiquity they were intimately linked [16]. A student of ancient pharmacology and botany can by deep exploration and skilled interpretation of the writings of these "Authorities" (Table 3.1) also hope for regular new discoveries. These ancient philosophers stressed on experience rather than acquiring competency with bookish knowledge. Totelin continues to observe "Plato in the *Phaedrus* (268c) stressed that nobody can become a physician by reading remedies in a book; Aristotle argued that nobody could learn medicine from books—books are only useful to those who are already experienced (*tois empeirois*), not to the ignorant (*tois anepistēmosin*) (*Nicomachean Ethics* 10.9, 1181b2–6); and Socrates in Xenophon's *Memorabilia* (4.2.10) mocked Euthydemus for attempting to become a physician through reading [17]." He goes on to analyze that the "experience" these authorities talk about is not only about first-hand observation of patients under treatment but also botanical expeditions to understand the life stages, seasonality and level of readiness to be harvested for therapy. It is also observed the pharmaco-botanists of yore laid a lot of importance on documentation. Galen, like other medical authors, often indicates that he has tried and tested a compound remedy, thus demonstrating that his knowledge was not merely bookish "Efficacy phrases," such as *expertum est* ("this has been tried") or *probatum est* ("this has been tested"), are common in medieval manuscripts, where they can be found either in the margins or in the main body of the text. While there may be a lot of merit in what the ancient texts prescribe, scientific temper demands empirical data and reproducibility.

Krishnaswamy and Kushalappa [18], while evaluating the micro-morphological structure of various parts of Andrographis serpyllifolia enumerated the impact of morphological adaptation of each organelle, its functional role in plant and potential bioactive accumulation (Table 3.2). Plant morphology is a function of the prevalent geo-ecological influences and the functional responses determine the phytochemical pathways and hence phytochemical accumulation.

Table 3.1 Treatises on Medicinal Plants Written by Various Authorities in History

S/N	Authority	Period	Name of Treatises	Salient Features and Authorship
1.	Theophrastus	Fourth century BCE	Enquiry into Plants	Information about pharmacologically active plants and the people who sold them, the pharmakopōlai (the drug sellers) (Amigues 1988–2006 [18]; 2010).
2.	Dioscorides	First century CE	Materia Medica	A five-book catalogue of pharmacologically active natural substances also contains some excellent botanical information (Wellmann 1906–1914 [19]; Riddle 1985 [20]; Beck 2005 [21]).
3.	Hippocrates	Fifth and fourth centuries BCE	Hippocratic Corpus	The gynaecology remedies, which contain a large number of recipes (Totelin 2009a [22]; Nutton 2013 [23]; Craik 2014 [24]).
4.	Pliny	AD 23–79	Naturalis Historia	It encompasses the fields of botany, zoology, astronomy, geology and mineralogy as well as the exploitation of those resources (French and Greenway 1986 [25]; Beagon 1992) [26].
5.	Oribasius	Fourth century CE	The medical encyclopedias	Scarborough (1984)
6.	Aetius and Alexander of Tralles	Both sixth century CE	The medical encyclopedias	
7.	Paul of Aegina	Seventh century CE	The medical encyclopedias	
8.	Galen	AD 129–200	The medical encyclopedias	The large pharmacological output
9.	Asclepiades the Pharmacist	First century CE 124 or 129–40 BC	External Ailments	Greek physician proposed a new theory of disease, based on the flow of atoms through pores in the body. Pioneer in molecular medicine.
10.	Titus Statilius Crito	Second century CE	Crito's Kosmetika, "On compound drugs according to places" "On compound drugs according to kinds"	Compilation of all cosmetic formulations of medicinal value rather than superficial
11.	Lucius Junius Moderatus Columella	4–c. 70 AD	Agronomist De Arboribus and De Agricultura	Diederich (2007) On farming or on agriculture
12.	Cato the Elder	160 BC	Agronomist De Agri Cultura	
13.	Crateuas the root cutter		Rhizotomikon (Root-cutting)	Personal physician to Mithridates VI Eupator, King of Pontus, compiled the first illustrated herbal which included, according to Pliny, "painted likenesses of plants."

This article incorporates text from sources, which are in the public domain.

Table 3.2 Correlation between Plant Morphology and Bioactive Accumulation in *A. serpyllifolia*

Morphological Feature	Description	Functional Aspects	Potential Bioactive Accumulation
Habit	Stout root stock	Storage of nutrients and water	Presence of carbohydrates
		Potential energy centers	Presence of starches
Juvenile shoots	Reddish purple shoots	Preservation of apical shoots	Presence of antioxidants (Flavonoids)
High soil temperatures	Soil surface and aerial microenvironment	Preservation of internal water	Presence of waxes, cuticle
			Terpenoids
Temperature, water stress	Succulent leaves reduced in size	Protection of cellular organelles from heat	Dihydroxy B-ring-substituted flavonoids
Stomata	Diacytic	Chloroplast protection (anti-oxidant activity)	Increased primary metabolism; photosynthesis
		Higher exchange of CO_2	Multiple glycolysis cycles, higher production of carbohydrates
Trichomes	Presence of short white streak-like deposits	Preservation of internal water, light reflectance	Could be biomineral deposits such as calcium or silica salts
	Relative abundance of unicellular long trichomes	Protection of cellular organelles from heat	Physical barrier for UV penetration
	Relative abundance of sessile trichomes on the abaxial laminar surfaces	Chloroplast protection (anti-oxidant activity) Protection from small pests on soil surface	Presence of flavonoids, unsaturated lipids
	Presence of glandular trichomes	Secretory function	Presence of terpenoids, low molecular weight volatiles, polysaccharides
Pollen	Yellow, round pollen with reticulate ornamentation	Protection of microspore from desiccation or inundation, UV radiation, heat	Sporopollenin comprised of long chain fatty acids, phenylpropanoids, phenolics, and traces of carotenoids.
Testa	Yellow, rugose with waxy fibrils	Protection of embryo from desiccation or inundation, UV radiation, heat	High molecular weight proteins, lipids, waxes

3.4 Pre-requisites for value-based bioprospecting

A robust bioprospecting experimental design may be built keeping the focus on a few key points:

- Pharmacological activity and target organ for which Herbal drugs are to be shortlisted
- Potential Candidate Herbs to evaluate based on Ethnomedicine, plant morphology and Doctrine of Signatures or Ayurvedic principles
- Adequate availability of herbal raw material
- Botanical, pharmacognostic, and phytochemical profiling standards
- Desirable Characters of Candidate molecules
- Choice of Pre-Set Bio assays to determine the efficacy of the candidate molecules

Once the Efficacy of the Phytocompound has been determined in the Assays, the bioprospecting lead molecule may be taken up further for animal and human trials and industrial scale manufacturing.

3.5 Biodiversity pool: the primary source

Fabricant and Farnsworth [27] identified a total of 122 compounds from only 94 ethnomedically qualified species of plants and 80% of these compounds were used for the same (or related) ethno-medical purposes, validating the traditional wisdom. In the first global assessment of the world's flora carried out by the Royal Botanic Gardens, Kew scientists have estimated that there are 390,900 plants known to science [28]. The study also found that 2,034 new plant species were discovered in 2015. In total, they now estimate that, excluding algae, mosses, liverworts, and hornworts, there are 390,900 plants, of which approximately 369,400 are flowering. "This is just scratching the surface. There are thousands out there that we don't know about," said Prof Kathy Willis, Director of Kew Botanic Gardens. Going by the ratio of plant to leads generated by Fabricant and Farnsworth (2000) the sheer number of potential hits possible (maybe approximately 490,000) is truly staggering!!

3.5.1 Herbal evaluation based on ethnomedicine, plant morphology, and doctrine of signatures or ayurvedic principles

The main challenge that bioprospectors face is to determine the best approach to discover "high net worth plants" that contain tangible quantities of potential drugs in a short period. Fabricant and Farnsworth [27] relate an episode when a Mexican physician presented small pieces (30 g) of the roots of a Mexican plant alleged to alleviate toothache pain, which created a profound local anaesthetic effect lasting for about 60 minutes. Evaluation of 50% ethanol extract in the acetic acid-induced writhing inhibition test in mice (i.g.).

A subfraction, showing one major spot following thin layer chromatography, gave an ED50 of 19.04 mg/kg (i.g.). Morphine showed an ED50 of 2.0 mg/kg (i.g.). Within 2 days a pure compound was isolated in high yield, identified and synthesized within 1 week. The pure compound was active in this assay, but 40% of the mice died within 40 minutes of administration at a dose of 40 mg/kg (i.g.). The ED50 of this compound was 6.98 mg/kg (i.g.). The plant was then identified as *Heliopsislongipes* (A. Gray) Blake, and the isolated bioactive compound was identified as the previously known isobutylamide, affinin (spilanthol). This demonstrated that the path to botanical drug discovery could be shorter if Ethnomedicine was the start line.

Koehn and Carter [29] have described the following unique phytochemical features of the potential compounds that could be isolated from natural products: (1) Greater number of chiral centers, (2) increased steric complexity, (3) higher number of oxygen atoms, (4) lower ratio of aromatic ring atoms to total heavy atoms, (5) higher number of solvated hydrogen bond donors and acceptors, (6) grater molecular rigidity, (7) broader distribution of molecular properties such as molecular mass, octane water partition coefficient, and diversity of ring systems. These observations help eliminate a large number of candidate molecules that don't fulfil the criteria and can be excluded from further testing.

3.5.2 Adequate and sustained availability of herbal raw material: baseline survey/inventory and occurrence of medicinal plants

There is need to conduct fresh survey in forest and other areas to find out different medicinal and aromatic plants growing therein along with their analytical characteristics such as frequency, density and species abundance. A plant inventory map of each unit area should follow the survey report indicating the presence of useful medicinal plant species. Accurate identification is pre-requisite for collection of authenticated material to avoid confusion due to species similarity biodiversity screening can be developed based on the taxonomic affiliations, genetic lineages, successful and considerable size of standing populations that indicate best evolutionary adaptive strategies of the medicinal plant in order to not only survive tough bio-geo-ecological conditions but also thrive in them.

Katy Moran [30] evaluated the pros and cons of various aspects of bioprospecting citing three case studies. The Tropical Botanic Garden and Research Institute (TBGRI), Trivandrum, was established by the government of Kerala, India, in 1979. In 1987, the TBGRI held an ethnobotanical field study in the forests of the Western Ghats in southwest India. The Kani, nomadic tribal forest dwellers of this region use a wild plant for energy that they called arogyapacha, and was identified as Trichopus zeylanicus by the TBGRI. It provided a lead to the development of the drug Jeevani (giver of life) after the TBGRI transferred the manufacturing licence to a major Ayurvedic drug company in India: The

Aryavaidya Pharmacy Coimbatore Ltd. licensed Jeevani as a tonic to bolster the immune system and provide energy for a fee of Rs 10 lakhs (approximately US$25,000). The TGBRI agreed to share 50% of the licence fee and the 2% royalty on profits with the Kani [31]. Dr. P. Pushpangadan, former director of the TBGRI, reports that arogyapacha, a perennial undergrowth, was cultivated to ensure a regular supply of the raw drug to the manufacturer [15]. But the TBGRI scientists learned that the medicinal qualities of the plant are lost, unless grown in natural forest settings. So the TBGRI organized 50 Kani families who live inside the forest to cultivate and pre-process the plant under the supervision of the TBGRI scientists. This plan, writes Pushpangadan, generates employment, as leaves of the plant are bought from the Kani who manage the semi-wild crop. Each family has one or two acres of arogyapacha under cultivation in the first year of the project and earns about Rs. 30,000 per acre. This income is expected to increase in subsequent years, as it is anticipated that the production of leaves will increase and last for 20–30 years. Since the plant can be grown only in the natural forest habitat to maintain its medicinal qualities, it seems the Kani are in an effective bargaining position to regulate and control its harvest. Pushpangadan describes Arogyapacha as forest friendly because it is grown under the shade of the natural forest canopy, but he does not refer to any sustainability studies on its management. Although it contributed no men, material or finances to the effort, Pushpangadan recently writes, that the State Forest Department now demands a share of the licence fee and royalties on the grounds that the plant material is collected within a forest area.

King et al. [32] while analysing different strategies employed to simultaneously address the need of the healthcare sector and the conservation of medicinal plants described the national and international efforts at *in situ* species conservation (within and outside of the conservation areas), *ex situ* species conservation (botanic gardens, seed banks, community gardens, herbal gardens), education and outreach programs, sustainable management of wild populations, and cultivation strategies for important medicinal plants. But, these strategies appear to overlook the role of plant—plant interactions, bio-geo-ecological interactions, inter species and inter genera synergies and antagonisms that contribute to the highly dynamic equation of bioactive stimulation, assimilation, and accumulation within the plant body. It is not surprising that the *Trichopus zeylanicus* cultivated by the Kani tribe failed to produce adequate phytochemical assay for commercial viability. The mono-cropping culture so successfully advocated for agricultural and horticultural crops largely fails to be just as rewarding in the case of medicinal plants. It is also true that there are exceptions and herbs like *Withania somnifera* (Ashwagandha) and *Gloriosa superba* respond very well to monocropping cultivation practices employed in India. But largely, it appears that a medicinal plant thrives under the challenge to its survival, the challenge being stress of any form: herbivory, harsh edaphic conditions, temperature/water/nutrient stress, competition, fungal/mycorrhizal associations, and competing genera/species.

Based on these observations, the author proposes the model of Biodiversity Plantation or Biodiversity cultivation. The intrinsic components of Biodiversity cultivation can be developed by paying keen attention to the natural dynamics of forest ecosystems: nutrient flow, soil microbe profiles, pollinator populations, pest-predator dynamics, microclimate build-up and maintenance, choice of plant genera and species that and known to best elicit the phytochemical responses. Do certain species level plant–plant interactions have a direct bearing on content, quality, and quantity of bioactives generated, which will impart instantaneous protection when required, and much like a surge of adrenaline would, in the animal kingdom? Can certain allelopathic stimuli elicit high secondary metabolite production? Investigating a deep mystery of mass deaths of Kudus of North Eastern African Savannah during severe drought, Wouter Van Hoven [33], a zoologist from Pretoria University, found that Acacia trees pass on an "alarm signal" to other trees when Kudus, a type of antelope, graze on their leaves. He says that acacias nibbled by antelope produce leaf tannins in quantities lethal to the grazers and emit ethylene into the air, which can travel up to 50 yards. The ethylene warns other trees of the impending danger, which then step up their own production of leaf tannin up to four times higher than normal, within just 5 to 10 minutes. Van Hoven made his discovery when asked to investigate the sudden death of some 3,000 South African antelope, called kudu, on game ranches in the Transvaal. He noticed that giraffe, roaming freely, browsed only on one acacia tree in ten, avoiding those trees which were downwind. Kudu, which are fenced in on the game ranches, have little other than acacia leaves to eat during the winter months. So they continue to graze until the tannin from the leaves sets off a lethal metabolic chain reaction in their bodies. Thus patterns of herbivory may contribute valuable clues regarding hyper-accumulation of specific bioactives in bioprospecting experiments. Dr. Edgar Warner assessed plants as "intelligent beings" that use a sophisticated system of perceptions and response: their perception of environment and their corresponding reactions, an adequate memory with a capacity to record, retain recall an event and respond accordingly. It may be worthwhile to study plant's chemical responses to herbivory threat, threat from man, allelopathy, stress, competition in a wholesome manner rather than in isolation.

3.5.3 Biodiversity plantation or cultivation

While traditional knowledge catalogues therapeutic value of various medicinal plants, there is poor knowledge of distribution and association of various medicinal plants in a given forest ecosystem. A Natural Ecosystem is developed with keen sustained observation of a standing forest population and planting carefully chosen multiple varieties in same unit area simulating natural plant associations such as thorny scrub jungle where the grasses and thorny species are dominant. The modern agricultural methods of monocropping systems are not the best management plan for medicinal plant cultivation. For one thing, the right agronomy for the relevant medicinal plant may not be available; and for another, the said medicinal plant may not prefer isolation from its naturally associated species. Often it has been proved that wild collections possess greater therapeutic value than cultivated ones.

3.5.3.1 Components of biodiversity plantation

The criteria of essentiality for establishment of a successful and economically rewarding biodiversity plantation include:

- Edaphic factors: Abiotic factor such as physical or chemical composition of the soil, soil profile, structure, porosity, soil moisture, soil air, soil enzymes found in a particular area. Biotic factors such as soil microflora: bacteria, fungi, algae viruses.
- Ecological events such as forest fires, drought, excess monsoon, extensive grazing.
- Plant combinations that facilitate full expression of the "inner potential" of each participating species.
- Optimum number of plant genera that can be successfully incorporated without crowding a unit area where consociation changes to competition.
- Microbial and fungal genetic resources.
- Economic returns.

A model simulating forest-derived combination can be adopted in a biodiversity plantation comprising medicinal plants growing in natural associations with little investment. Trees, shrubs, and undergrowth can be incorporated into a multitier, multicropping system to work around seasonality and maximize land utilisation and productivity per unit area. It needs a multidisciplinary collaboration of botanists, geologists, phytochemists, and agronomists to identify the right plant genera/species combination that not only creates a winning combination but also covers the aspect of commercial viability, to choose participating species with high therapeutic or curative value and sustained demand. Different parameters such as weather patterns traced through at least a three-year period, recording the soil surface temperature, residual moisture, ambient humidity, frequency of plant occurrence along with other associated herbs, grasses and shrubs need to be investigated. Soil analysis to determine its carbon, macronutrient and micronutrient profile, relative abundance of any mineral or salt and pH, can help in devising the ideal agronomic practices. The soil texture, pH, organic matter, nutritional profile and Ca-Mg ratios determine bioavailability of nutrients in the soil and hence successful establishment of a few species. Soil enzymes are released by microorganism, soil animals like earthworms and plant roots, and maintain soil fertility by catalysing biological reactions in soil. The common soil enzymes influencing soil texture and nutrient status are amylases, catalases, invertases, dehydrogenases, phenol oxidases, glycerophosphatases, and ureases [34]. For instance, a very alkaline soil that lacks dehydrogenases may be a limiting edaphic factor restricting the variety of plants growing in a region. Picking up clues about the combination of plant species that may contribute through positive synergistic association in the example above, *Hemidesmes indicus* (Ananthamool) is a slow-growing small straggly creeper, and Pregnane glycosides and Sitosterol [35,36] in the roots are highly valuable as blood purifier. It is also on the IUCN "Rare or depleted in Western Ghats due to over exploitation" list. *Andrographis serpyllifolia* with multiple medicinal properties is a slow-growing geophyte, *Tylophora asthamatica* is a relatively fast-growing creeper with a sustained demand. These three varieties can be intercropped with deciduous trees and shrubs, which insulate the undergrowth from pollution. This combination may be considered as a candidate-group for Biodiversity Plantation trials in arid zones.

3.6 Conclusion

The plant body appears to be in a state of constant heightened awareness and vigil about minute changes in its micro- and macro-environments, which may be the reason why despite innumerable challenges, the plant kingdom still survives successfully. Biodiversity, when left alone, flourishes as a single conscious mass of various beings but the threat from man-made challenges such as habitat destruction, industrial effluents, and other forms of pollution are tightening the noose quite rapidly to choke the life breath away. Bioprospecters should ideally consider compiling "pre-bioprospecting intelligence" to evaluate the key botanical signatures, ecological factors, composition of highly evolved species with ingenious survival strategies, before embarking on harnessing bioactives from nature. This approach may reduce time, effort, and expense by eliminating testing of potentially non-productive species in a population.

References

1. Scott P. (2001). Bioprospecting as a conservation tool: History and background From Crossing Boundaries in Park Management. In D. Harmon (Eds.). *Proceedings of the 11th Conference on Research and Resource Management in Parks and on Public Lands*. The George Wright Society, Hancock, MI.
2. Valiathan M. S. (2003). *The Legacy of Caraka*. Orient Blackswan, Hyderabad, India.
3. Mon G. L. (2017). http://www.natural-wonder-pets.com/do-wild-animals-heal-themselves.html.
4. Inglis-Arkell E. (2014). Elephants might be able to self-medicate to induce labor. Available at http://io9.gizmodo.com/elephants-might-be-able-to-self-medicate-to-induce-labo-1611904103.
5. Burkil H. M. (2004). *The Useful Plants of West Tropical Africa*. Royal Botanic Gardens Kew, Richmond, UK.
6. Nguelefack T. B., Watcho P., Wansi S. L., Mbonuh N. M., Ngamga D., Tane P. et al. (2005). The antiulcer effects of the methanol extract of the leaves of *Aspilia africana* (Asteraceae) in rats. *African Journal for Traditional Complementry and Alternative Medicine*, 2, 233–237.

7. Farombi E. O. and Owoeye O. (2011.) Antioxidative and chemopreventive properties of *Vernonia amygdalina* and *Garcinia biflavonoid*. *International Journal for Environmental Research and Public Health*, 8(6), 2533–2555.
8. Gran G. (2014) The Karma of plants. Available at http://yogachicago.com/2014/01/the-karma-of-plants/.
9. Cassaro R. (2011) The ancient "Doctrine of Signatures" suppressed by the establishment. Available at https://www.richardcassaro.com/the-ancient-doctrine-of-signatures-suppressed-bythe-elite.
10. Wood M. (2011). The Doctrine of Signatures—Natura Sophia. Available http://www.naturasophia. com/Signatures.html.
11. Steinauer G. (2012). Healing plants of the Prairie. Available http://nativeplants. ku.edu/wp-content/uploads/2012/08/medicinal-plant-final-pdf.pdf.
12. Tierra M. (1983). The way of herbs. Available at https://www.amazon.com/Way-Herbs-Michael-Tierra/dp/0671466860.
13. Strobel G. and Daisy B. (2003). Bioprospecting for microbial endophytes and their natural products. *Microbiology and Molecular Biology Reviews*, 67, 4491–502.
14. Strobel G. A., Li J. Y., Sugawara F., Koshino H., Harper J. and Hess W. M. (1999). Oocydin A, a chlorinated macrocyclic lactone with potent anti-oomycete activity from *Serratia marcescens*. *Microbiology* 145, 3557–3564.
15. Cragg G. M., Boyd M. R., Cardellina J. H., Newman D. J., Snader, K. M. and McCloud T. G. (1994). Ethnobotany and drug discovery: The experience of the US National Cancer Institute. *Ciba Foundation Symposium* 185, 178–190.
16. Bills G., Dombrowski A., Pelaez F., Polishook J. and An Z. (2002). Recent and future discoveries of pharmacologically active metabolites from tropical fungi. In R. Watling, J. C. Frankland, A. M. Ainsworth, S. Issac, and C. H. Robinson. (Eds.). *Tropical Mycology: Micromycetes*. CABI Publishing, New York.
17. Totelin L. M. V. (2016). Subject: Classical studies, ancient science and medicine. doi:10.1093/oxfordhb/9780199935390.013.94.
18. Krishnaswamy S. and Kushalappa, A.B. (2017). Correlative evaluation of the impact of adaptive plant morphology on bioactive accumulation based on micro-morphological studies in *Andrographis serpyllifolia* (Rottler ex Vahl) wight. *Notulae Scientia Biologicae*, 9(2), 263–273.
19. Amigues, S. (1988–2006). *Théophraste: Recherches sur les plantes. 5 vols. Texte établi et traduit par S. Amigues*. Les Belles Lettres, Paris, France.
20. Wellmann, M. (1906–1914). *Pedanii Dioscuridi Anazarbei De materia medica libri quinque. Edidit M. Wellmann. 3vols*. Weidmann, Berlin, Germany.
21. Riddle, J. M. (1985). *Dioscorides on Pharmacy and Medicine*. University of Texas Press, Austin, TX.
22. Beck, L. Y. (2005). *Pedanius Dioscorides of Anazarbus "De materia medica"*. *Translated by L. Y.* Hildesheim, B. Olms-Weidmann, Hildesheim, Germany.
23. Totelin L. M. V. (2009). *Hippocratic Recipes: Oral and Written Transmission of Pharmacological Knowledge in Fifth- and Fourth-Century Greece*. Brill, Leiden, the Netherlands.
24. Nutton V. (2103). *Ancient Medicine*, 2nd ed. Routledge, London, UK and New York.
25. Craik E.M. (2006). *Two Hippocratic Treatises on Sight and on Anatomy*. Brill, Leiden, the Netherlands.
26. French R. and Greenaway F. (1986). *Science in the Early Roman Empire: Pliny the Elder, His Sources and Influence*. Croom Helm, London, UK.
27. Beagon M. (1992). *Roman Nature: The Thought of Pliny the Elder*. Oxford University Press, Oxford.
28. Fabricant D.S. and Farnsworth N.R. (2001). The value of plants used in traditional medicine for drug discovery. *Environmental Health Perspectives*, 109, 69–75.
29. Morelle R. (2016). Kew report makes new tally for number of world's plants. Science & Environment World Asia UK Business Tech Science Magazine, *BBC News*.

30. Koehn F. E. and Carter G. T. (2005). The evolving role of natural products in drug discovery. *Nature Reviews Drug Discovery*, 4(3), 206–220.
31. Moran K. (2000) Bioprospecting: Lessons from benefit-sharing experiences. *International Journal of Biotechnology*, 2(1–3), 132–144.
32. Anuradha R. V. (1998). Sharing with the Kanis: A case study from Kerala, India, Secretariat to the Convention on Biological Diversity, Fourth Meeting of the COP, Bratislava, Slovakia.
33. King S., Meza E., Carlson T., Chinnock J., Moran K., and Borges J. (1999). Issues in the commercialization of medicinal plants. *Herbalgram*, 47, 46–51.
34. Hoven W. V. (1990). Clearing the airways. T. Beardsley (Eds.). *Scientific American* 263(6), 28–33.
35. Edaphic factors- soil profile, structure, porosity, soil moisture, soil air, University of Maryland School of Medicine. Baltimore, MD. Available at https://www.slideshare.net/siddumn/edaphic-factors.
36. Chatterjee R. and Bhattacharya B. (1995). A note on the isolation of beta-sitoserol from *Hemidesmus indicus*. *Journal of the Indian Chemical Society*, 32, 485–486.

4

Role of Reverse Pharmacology, Translational Medicine, and Indian System of Medicine in BDPs Development

Jayant N. Lokhande, Sonali Lokhande, and Ganesh Shinde

Contents

The average cost and time of discovering, developing, and launching a new drug by the pharmaceutical industry is increasing without an expected analogous increase in the number of newer, safer, and better drugs. The number of approvals for new drugs has declined dramatically from 53 in the year 1996 to just 17 in 2007 [1]. There seems to be "Blockbuster Drugs" as a myth nowadays and also the more stringent regulatory and vigilance process by regulatory bodies are intensifying risks and completion time for drug discovery and development projects.

Notwithstanding the dramatic increase of global spending on drug discovery and development, the approval rate for new drugs is declining, due chiefly to toxicity and undesirable side effects. In fact, lack of efficacy and toxicity are the two main causes for attrition in drug development, accounting for 60% of the drug failures [2].

The industry is facing a major challenge to sustain and grow, which is resulting in many mergers, acquisitions or closures [3]. In current circumstances techniques like Reverse Pharmacology and Translational Medicine can be helpful in designing new drugs and cut down the risk of drug failures.

4.1 What is reverse pharmacology

Reverse Pharmacology is an integrated science as it consists of experimental traditional medicine, collecting hits from it and then translating hits into leads and or new drugs with clinical studies.

Reverse Pharmacology primarily involves four stages:

- Empirical Evidence, Secondary Data Research, and Pharmaco-epidemiology to create Hits
- In vitro and in vivo studies to transform Hits in to Leads with initial proof of concept with Systems Biology Approach
- Establishment of Safety and Efficacy of "Leads—Prospective Drug Candidates" with possible action mechanism with clinical trials
- Intellectual Property Generation and Regulatory Compliance

Reverse Pharmacology can establish integrative collaboration in between modern and ancient medical science. This science utilizes traditional knowledge of medicines to rediscover drugs and is also called as a pharma-viaduct in between bed and bench experiments.

Reverse Pharmacology when coupled with Observational and or Empirical Therapeutics, Pharmaco-epidemiology and Systems Biology would enhance and complement the new drug discovery process. The high cost of modern drugs may then hopefully be reduced. The insights obtained in non-drug interventions also can reduce the out-of-control health care budgets.

Reverse Pharmacology as a term has two nuances. The first is the path of pharmacology from the bedside observations to bench experiments. The second is the search of drug-like molecules (endogenous or exogenous) dock-in with new macromolecules discovered through genomics and proteomics [4] (Figure 4.1).

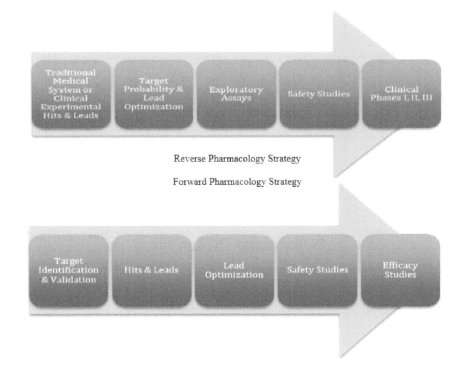

Figure 4.1 *Reverse and forward pharmacology strategy.*

4.2 Role of reverse pharmacology and pharmacognosy in BDP development

The Plant Kingdom has contributed to Modern Pharmacology in great respect; so far, 60% of anticancer and 75% of anti-infective drugs approved from 1981 to 2002 could be originating from natural products [5].

As Ethno-medicinal Systems are being practiced since from ages, not only its safety is being proved but also its evidence based therapeutic sustainability. Many botanical-originated medicines in ethnomedicine can prove out to be workable scaffolds for rational drug design. The altogether new and/or discreet therapeutic applications can be sought through ancient chemical entities used in ethno medicine.

Few decades ago, Hoechst and Central Drug Research Institute (CDRI), CSIR, Lucknow isolated phytochemicals like Forskolin from Coleus forskholii and Coleonol from Stephania glabra, respectively. Currently Forskolin is now being rediscovered as adenylate cyclase for obesity prevention and Coleonol as nitric oxide activators for atherosclerosis prevention [6,7].

The Reverse Pharmacology schema is illustrated in Figure 4.2.

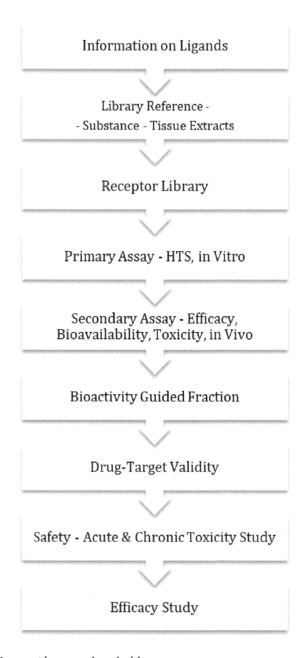

Figure 4.2 *Reverse pharmacology ladder.*

There are various tools and techniques can be employed in exploring botanicals information to understand its Reverse Pharmacology potential.

4.2.1 Current database for natural compounds

There are various databases available to explore preliminary data on Botanicals, its compounds, chemical structures, molecular structures, and its Past Therapeutics History. Virtual screening workflows usually involve docking a compound library into the binding site of a target receptor and using scoring functions and binding free energy calculations to identify putative binders.

  Natural compounds that appear in the published literature and compounds found in commercial databases forms the structural database called as Virtual Chemical Database (VCDB). The sources of these compounds are available, and frequently the method applied for their extraction is also described. CONCORD is a VCDB that contains more than 100,000 natural compounds, along with their 3D coordinates. Chemical diversity of the compound is the final criterion for the selection of compounds for virtual screening [8].

  https://pubchem.ncbi.nlm.nih.gov – Contains chemical structures and biological properties of molecules including small molecules and siRNA reagents. PubChem consists of three interconnected databases: Substance, BioAssay, and Compound. The database also provides a suite of web-based bioactivity analysis tools allowing to download and search individual test results, compare biological activity data from multiple screenings, examine target selectivity or explore structure–activity relationships for compounds of interest.

  https://gnps.ucsd.edu/ProteoSAFe/static/gnps-splash.jsp – Global Natural Products Social Molecular Networking (GNPS) web-platform provides public data set deposition and/or retrieval through the Mass Spectrometry Interactive Virtual Environment (MassIVE) data repository. The GNPS analysis infrastructure further enables online dereplication, automated molecular networking analysis, and crowd sourced MS/MS spectrum curation. Each data set added to the GNPS repository is automatically re-analyzed in the next monthly cycle of continuous identification.

  http://www.megabionet.org/tcmid/ – Traditional Chinese Medicine Integrated Database (TCMID) is a comprehensive database to provide information and bridge the gap between Traditional Chinese Medicine and modern life sciences. It has collected information on all respects of TCM including formulae, herbs, and herbal ingredients.

  http://bioinf-applied.charite.de/supernatural_new/index.php – Super Natural II, a database of natural products. It contains 325,508 natural compounds (NCs), including information about the corresponding 2D structures, physicochemical properties, predicted toxicity class, and potential vendors. Natural products are small compounds synthesized by living organisms. The chemical diversity of these molecules is tremendous and offers inspiration for innovations in medicine, nutrition, agrochemical research and life sciences. Most of the currently used cosmetics and drugs are either natural products or close derivatives thereof.

http://bioinformatics.psb.ugent.be/triforc/#/home – TriForC is an acronym for the project "A pipeline for the discovery, sustainable production and commercial utilisation of known and novel high-value triterpenes with new or superior biological activities." This is an EU-funded collaborative project on establishing an integrative and innovative pipeline for the exploitation of plant triterpenes.

http://crdd.osdd.net/raghava/npact/ – NPACT is a curated database of plant-derived natural compounds that exhibit anti-cancerous activity. It contains 1,574 entries and each record provides information on their structure, properties (physical, elemental, and topological), cancer type, cell lines, inhibitory values (IC50, ED50, EC50, GI50), molecular targets, commercial suppliers, and drug likeness of compounds. NPACT concentrates on anti-cancer natural compounds found in plants only. NPACT is unique in providing bioactivities of these natural compounds against different cancer cell lines and their molecular target.

http://bidd2.nus.edu.sg/NPASS/ – The NPASS database in the current version (V1.0) provides 35,032 unique natural products isolated from 25,041 source organisms and together with 446,552 activity records on 5,863 targets. The NPASS will be regularly updated to include more natural products, which have both species source and quantitative activity data available from recent publications.

http://www.cemtdd.com – Contains about 621 herbs, 4,060 compounds, 2,163 targets, and 210 diseases, among which most of herbs can be applied into gerontology therapy including inflammation, cardiovascular disease and neurodegenerative disease. CEMTDD displays networks for intricate relationships between Chinese Ethnic Minority Traditional Drugs (CEMTDs) and treated diseases, as well as the interrelations between active compounds and action targets, which may shed new light on the combination therapy of CEMTDs and further understanding of their herb molecular mechanisms for better modernized utilizations of CEMTDs, especially in gerontology.

http://nubbe.iq.unesp.br/portal/nubbedb.html – Collects classes of secondary metabolites and derivatives from the biodiversity of Brazil. NuBBEDB is a database of compounds from species of the Brazilian biodiversity, especially from the two main biomes Cerrado and Atlantic Forest. It provides botanical, chemical, pharmacological, and toxicological information and includes a variety of information for each compound including chemical class and name, code, molecular formula, mass, and source. A web-based search tool allows users to search for compounds by property, chemical structure, or a combination of criteria.

http://c13.fundacionusal.es – This database contains (13)C spectral information of over 6,000 natural compounds, which allows for fast identifications of known compounds present in the crude extracts and provides insight into the structural elucidation of unknown compounds.

https://sancdb.rubi.ru.ac.za – SANCDB is currently the only web-based NP database in Africa. It aims to provide a useful resource for the in silico screening of South African NPs for drug discovery purposes. There is a submission pipeline to allow growth by entries made from researchers that supports the database. As such, SANCDB is the starting point of a platform for a community-driven, curated database to further natural products research in South Africa.

http://ab-openlab.csir.res.in/biophytmol/ – BioPhytMol is a manually curated drug discovery community resource on anti-mycobacterial phytomolecules and plant extracts. Currently, the resource holds a total of 2,582 entries, including 188 plant families (composed of 692 genera and 808 species) and 633 active compounds and plant extracts identified against 25 target mycobacteria.

http://mesh.tcm.microbioinformatics.org – Database on Traditional Chinese Medicine preparations.

http://silver.sejong.ac.kr/npcare/ – Natural Products CARE (NPCARE), a database for Natural Products-CAncer gene REgulation, provides the level of gene expression and the inhibition of cancer cells in various cancer types by the effect of extract and Natural compounds from more than 2,000 native species including plants, marine species and microorganisms. 700 genes and 1,100 cancer cell lines annotated by expert enable users to gain insight into finding potential anti-cancer drugs and understanding the mechanism of Natural products for cancer treatment.

http://www.3dmet.dna.affrc.go.jp – A database of 3D structures of natural metabolites has been developed called 3DMET. During the process of structure conversion from 2D to 3D, we found many structures were falsely converted at chiral atoms and bonds. The current version includes most of the natural products of the KEGG COMPOUND collection [http://www.genome.jp/kegg/compound/] and is searchable by string, value range, and substructure.

http://cadd.gdhtcm.com:2180/PDTCM/ – Provides users a psoriasis database of Traditional Chinese Medicine with systems pharmacology-based methods. PDTCM comprises eight data entities including herbal formulas, Traditional Chinese medicine (TCM), source plants/animals, molecules, target proteins, docking results between all molecules and target proteins, diseases, and clinical biomarkers. It also offers clues and ideas in exploring perspective and effective TCM for psoriasis.

AfroDb – AfroDb represents the largest "drug-like" and diverse collection of 3D structures of NPs covering the geographical region of the entire African continent, which is readily available for download and use in virtual screening campaigns. The availability of 3D structures of the compounds to be used for docking is of utmost importance. Therefore, the availability of such structures within AfroDb, as well as their calculated physico-chemical properties and indicators of drug-likeness within this newly developed database will facilitate the drug discovery process from leads that have been identified from African medicinal plants [9].

http://african-compounds.org/nanpdb/ – Data currently comprising 641 data sources (covering the period from 1962 to 2016) and was derived from literature assembled from the main stream natural product journals as well as MSc and PhD theses in some African university libraries and selected searches in local African journals. The data covers compounds isolated mainly from plants, with contributions from endophytes, animals (e.g., corals), fungi, and some bacteria sources. In addition to names and molecular structures of the compounds, information about source organisms, references, biological activities and modes of action (e.g., antimalarial, anticancer, cytotoxic, etc.) are also mentioned. Data can be accessed via queries on compound names, chemical structures, organisms, or keywords. All compound structures can be downloaded entirely from the website and can be applied for in-silico screenings for identifying new active molecules with undiscovered properties.

http://tcm.lifescience.ntu.edu.tw – TCMGeneDIT is a database system providing association information about Traditional Chinese Medicines (TCMs), genes, diseases, TCM effects, and TCM ingredients automatically mined from vast amount of biomedical literature. Integrated protein-protein interaction and biological pathways information collected from public databases are also available. In addition, the transitive relationships among genes, TCMs, and diseases could be inferred through the shared intermediates. Furthermore, TCMGeneDIT is useful in deducing possible synergistic or antagonistic contributions of the prescription components to the overall therapeutic effects. TCMGeneDIT is a unique database of various association information about TCMs. The database integrating TCMs with life sciences and biomedical studies would facilitate the modern clinical research and the understanding of therapeutic mechanisms of TCMs and gene regulations.

http://informatics.kiom.re.kr/compound/ – TM-MC provides information on the constituent compounds of medicinal plant materials in Northeast Asia traditional medicine. Information on the constituent compounds of these medicinal materials was extracted manually from chromatography articles searched on MEDLINE and PubMed Central. In the field of Northeast Asian traditional medicine, many databases exist that provide the constituent compounds of medicinal materials. However, some databases are not open to the public or provide much overlapping information. Moreover, although there exist many databases, it is difficult to decide which database to use because the sources of the information are unclear. Through this database the aim is to list Northeast Asian (Korea, China, and Japan) national pharmacopoeias medicinal plants material and or medicinal extracts with which users can easily check relevant information with the links to articles.

http://tcm.cmu.edu.tw – Provides chemical compounds isolated from Chinese traditional herbs. The Traditional Chinese Medicine Drugs Information System is a 3D structure database. This system enables (i) easy retrieval molecular information from a remote database, (ii) simple query as well as complex logic query on almost every field of a record, (iii) incremental query upon previous search, (iv) easy substructure query, (v) various display modes with many fine-tuned options for 3D structure visualization, (vi) calculation of bond distance, torsion angle, and bond angle on a 3D structure, (vii) batch 3D structures output to integrate with other commercial molecular simulation or drug design software.

http://nipgr.res.in/Essoildb/ – Provides a library compiling plant essential oils. EssOilDB provides detailed information such as family information about the source plant, origin of the material, and complete literature citation. The database also includes experimentally recorded essential oil profiles from several published reports. Users can combine different parameters as environmental conditions or geographical distribution in a single search to highlight the role played by each of them in the essential oil composition of the studied plant.

http://carotenoiddb.jp – The Carotenoids Database currently provides information on 1,175 natural carotenoids with 700 source organisms. Data on their structures and source organisms were obtained from the latest available original papers. The original chemical fingerprints newly investigated and described here make it easy to classify chemical modification patterns in carotenoid structures, to similarity search, and to predict some of the biological functions of carotenoids such as provitamin A, membrane stabilizers, odorous substances, allelochemicals, antiproliferative activity against cancer cells, and reverse MDR activity at the present time. The goal is to understand how organisms are related via carotenoids, either evolutionarily or symbiotically or in food chains through natural histories.

http://ayurveda.pharmaexpert.ru – Provides a resource for ayurvedic phytocomponents of medicinal plants. Ayurveda contains more than 50 medicinal plants, 1,900 phytocomponents, and around 950 natural compounds. The components selected all answer to these criteria: Ayurvedic or traditional medicinal use; adequately explored for phytochemical analysis; unexplored for pleiotropic pharmacological studies. The database offers a way to explore the mechanisms of action and pharmacological effects of individual components or combinations of medicinal plants.

https://mccordresearch.com.au – Gathers compounds identified from Olea europaea. OliveNetTM describes more than 600 compounds associated with O. europaea, including about 200 phenolic compounds. Users can examine their chemical structure for structure-activity relationship analyses and for potential use in virtual screening. The database can serve as a resource for those interested in compounds found within the various matrices of the olive, the associated analytical techniques used in their identification and/or isolation.

https://cb.imsc.res.in/imppat/ – Gathers information about traditional Indian medicine. IMPPAT is a manually curated database from scientific literature that aims to support in silico drug discovery. It assembles over 1,700 Indian medicinal plants, more than 9,500 phytochemical constituents, and 1,100 therapeutic uses. Searches can be made by phytochemical composition, therapeutic use and traditional medicinal formulation.

http://cosylab.iiitd.edu.in/spicerx/ – Provides a repository for health effects of culinary spices and herbs. SpiceRx collects tripartite associations between culinary spices/herbs, diseases, and phytochemicals. This database gathers up to 180 culinary spices and herbs and 150 spices are associated with more than 840 unique disease-specific Medical Subject Headings (MeSH) IDs. It offers a disease search function integrated with the hierarchical organization of MeSH disease terms.

http://www.nadi-discovery.com – Natural Based Discovery resource that has been developed with an intended aim to capture the plethora of information about Malaysia fauna and flora. It consists of NADI-CHEM, a Natural Product 3D Chemical Structure Database of Malaysian natural products and NADI-Herb as well as NADI-MEPS, databases of Herbal Monograph to provide information for natural-based drug discovery. NADI-CHEM has been developed in a format ready for virtual screening and with NADI-VISAGE, molecular docking can be performed to identify possible natural compounds with high biological activity in the development of a lead compound for a drug discovery program.

http://dmnp.chemnetbase.com/faces/chemical/ChemicalSearch.xhtml – A comprehensive database containing over 30,000 compounds. DMNP is a subset of the Dictionary of Natural Products (DNP) database. DNP is an ongoing project based on a 25-year review of the natural product literature. For the present project, the subset of DNP entries referring to marine natural products were carefully checked and reviewed and enhanced with a considerable amount of additional information relating to their natural occurrence. Several careful reviews were also carried out to ensure that the coverage of marine natural products in the finished publication was as complete as possible.

http://www.cbrg.riken.jp/npedia/index.php?LANG=en – Allows to search chemical compounds for Natural Products Encyclopedia of RIKEN. RIKEN NPEdia includes a search engine. It offers users to search information by two ways: a simple keyword search, and an advanced search (the research is customizable according to ID, general information, source organism, biological activity, or basic property).

http://pubs.rsc.org/marinlit/ – MarinLit is a database dedicated to marine natural products research. Professors John Blunt and Murray Munro at the University of Canterbury, New Zealand established the database in the 1970s. It was designed as an in-house system to fulfill the needs of the University of Canterbury Marine Group and has evolved to contain unique searchable features and powerful dereplication tools. The extremely comprehensive range of data contained along with these powerful features makes MarinLit the database of choice for marine natural products researchers.

4.2.2 Virtual screening

Virtual screening can be defined as a set of computational methods that analyzes large databases or collections of compounds to identify potential hit candidates.

The basic goal of virtual screening is the reduction of the enormous virtual chemical space of small organic molecules, to synthesize and screen against a specific target protein, to a manageable number of compounds that exhibit the highest chance to lead to a drug candidate.

There are two methods for virtual screening:

- Structure-Based Virtual Screening (SBVS) – This requires knowledge of the 3D structure of the target and the ligand. The techniques of de novo design and docking are involved for generating new ligands or adjusting ligands in the active site of the target respectively.
- Ligand-Based Virtual Screening (LBVS) – i.e., physicochemical properties (1D), Fragmental Description (2D) and Pharmacophores (3D data) that are techniques of Quantitative Structure-Activity Relationship (QSAR).

Structure-based virtual screening (SBVS) applies different modeling techniques to mimic the binding interaction of a ligand to a bio-molecular target and Ligand-based virtual screening (LBVS) uses 2D or 3D similarity searches between large compound databases and known actives. Hence, the biggest difference between LBVS and SBVS is that the latter requires structural information for the target, usually obtained from X-ray Crystallography or Nuclear Magnetic Resonance (NMR). The target database is composed of 3D protein structures, determined by X-ray crystallography or by homology modeling. The majority of the structures are from humans, although it also sometimes contains proteins from other sources (e.g., viruses, bacteria) [10]. If Structure-based information does not exist, e.g., membrane receptors such as GPCR, then this information can be mimicked with their homology models. Although docking is arguably the most widely used approach in early phase drug discovery and LBVS methods in general yield a higher fraction of potent hits [11,12].

Structure-based screening begins with target and the compound library preparation, then running the actual docking algorithms, post-processing and ranking the results for bioassays by a pre-defined scoring function [13]. The challenge in docking software is to consider protein flexibility. These macromolecules are obviously not static objects. The conformational changes are often key elements in ligand binding. Using multiple high-quality static receptor conformations as snapshots in docking runs and selecting the highest scoring conformation for further investigation is one way to tackle the problem. In Soft Docking (Molecular Dynamics Simulations) protein and ligand interaction is allowed to change dynamically [14–19].

(D.J. Abraham, Virtual Screening, Burger's Medicinal Chemistry Drug Discovery, 1(6), 243–280 (2003))

Water molecules can significantly affect ligand binding through the formation of hydrogen bonds and can contribute to both the enthalpy and entropy of the binding. In general, the thermodynamics of ligand-receptor interactions are still treated similarly to how molecular reactions work, and often times this is not the optimal way to approach the problem. As far as accounting for water, several approaches have been developed recently, which are complementary to the experimental information from X-ray and NMR spectroscopy [20–23].

Neural networks, support vector machines and the random forest techniques can describe the nonlinear dependence of the ligand-target interactions during binding without taking solvation and entropic effects into account. The structural interaction fingerprint (SIFt) method, which uses the 3D structure of the protein-ligand complex to generate a 1D binary fingerprint, is also important. This fingerprint then is used to characterize ligand poses derived from the docking procedure and compare to the native substrate's interaction map [24,25].

The ligand-based screenings do not take the target structure directly into account. LBVS techniques are based on the assumption that compounds with a similar topology have similar biological activity. There are many ways to define molecular similarity; ligand-based screening in drug discovery typically uses topology-based descriptors involving the pharmacophoric sites of the molecules. The descriptors of the known active molecules and the potential hit molecules are compared using pre-defined mathematical expressions (metrics) to quantify molecular similarity. These approaches essentially neglect any information about the target biomolecule as well as the 3D structure of the ligand compounds. Nevertheless, they are very efficient and are often applied in combination with structure-based approaches to identify potential bioactive hits that can then be fed into docking experiments [26].

Besides the structure and traditional ligand-based methods there is a third possible approach to predict the bioactivity of molecules in a virtual chemical space. Methods that belong to this third branch can be thought of as extensions of the ligand-based approach with the major difference that instead of considering only the molecular topology, they create or consider 3D coordinates of both the active and the potential lead molecules for the similarity comparison, and then estimate the 3D shape similarity of these molecules. These algorithms are called shape methods, although just like in the traditional ligand-based algorithms, a number of different ways to generate 3D similarity measures exist [26].

The combination of structure and ligand-based strategies Hybrid Approach, in which ligand- and structure-based applications are truly molded together (protein-ligand pharmacophores) and integrated into one standalone technique to enhance accuracy and performance, can also be employed in certain cases of Botanical Drug Products Discovery [27,28].

By combining advances in chemoinformatics and structural biology, it might be possible to rationally design the next generation of promiscuous drugs with polypharmacology [29,30].

4.3 Polypharmacology based translational drug discovery model in BDP

Botanical mixtures that contain hundreds of potentially bioactive natural products often do not reveal the desired molecular mechanisms of action. A major bottleneck in botanical drug research is the identification of the molecular targets of all bioavailable compounds within the extract (i.e., the overall mode of action). In many cases, no feasible mode of action can be elucidated by using conventional biochemical methodologies [31].

Botanicals do not produce secondary metabolites to benefit mammals but produce them to potentially address the diverse ecological pressures, such as microbial or predator attack. Plants often respond to a stressor by increasing the biosynthesis of different classes of molecules rather than just an individual secondary metabolite and their mixtures may simply affect the solubility and distribution of the potentially active constituents [32,33].

The bioactive compounds are very rarely produced alone, but almost always occur in mixtures with other potentially bioactive secondary metabolites with many physiological mediators (i.e., amines, fatty acid metabolites, steroids, etc.) that act by simultaneously targeting different sites [34]. The secondary metabolic pathways, which are turned on in response to specific cues and make natural products, typically make a variety of products. Some pathways make only one or two products, and some make more than 100. Thus, primary metabolic pathways are target-oriented, whereas secondary pathways are diversity-oriented [35,36].

In numerous Botanical Drug-Target relationships, no linear mode of action can be determined but then how Botanical Drugs those are in polycomposites forms exhibit therapeutic actions? The Network Pharmacology or Poly Pharmacology can lead to an answer. It would be possible that these poly molecular composites are targeting disease causing different protein networks at a time as disease involves more than one protein interaction cascade. Naturally there are functional relationships (systems biology) in living organisms, and genes and proteins are not isolated in space and time.

Integrating network biology and polypharmacology holds the promise of expanding the current concepts on druggable targets and may also help to understand the pharmacological action of botanical drugs [37]. Instead of thinking in terms of individual protein targets, there is a need to portray ligands as having multiple targets which are connected to other targets or functional networks. Molecules are connected to each other in an ordered manner defined by binding interactions in time and space rather than in a loose hodgepodge. There is good evidence that many anti-inflammatory plant-natural products act by binding to and blocking proteins, and this may be the reason why there are so many anti-inflammatory medicinal plants [34,38].

In Sativex (GW Pharmaceuticals Branded Drug) indicated for various neurological pains is characterized on multiple markers like, (* marked compounds are characterized and controlled in Botanical Drug Product specifications and # marked compounds are controlled and characterized in Botanical Drug Substance)

- Principal Cannabinoids: THC*27 mg/mL, CBD*25 mg/mL
- Minor Cannabinoids: CBC*, CBG*, CBN*, THC-V*, CBD-V*, THCA*, CBDA*, CBO#, CBE#, CBC-V, CBL
- Terpenes: – trans-caryophyllene#, α-caryophyllene#, caryophyllene oxide, α-pinene, β-pinene, terpinolene, myrcene, limonene, linalool, cis-nerolidol, trans-nerolidol, phytol, squalene
- Carotenoids: β-carotene#
- Fatty Acids: Linoleic acid, Palmitoleic acid, Linolenic acid, Palmitic acid, Oleic acid, Stearic acid, Myristic acid, Arachidic acid and Behenic acid
- Sterols: B-sitosterol, campesterol, stigmasterol
- Vitamins: Vitamin E
- Triglycerides: Trilinolenin, Trilinolein.......

Another hindering factor in Botanical Drug Products research is obscure pharmacodynamics of potentially bioactive natural products in drug mixtures. It is very important to emphasize time and again that bioactive natural products do not necessarily possess optimal ADME (administration – distribution – metabolism – excretion) properties in mammals because the latter have evolved detoxifying mechanisms for many common chemical scaffolds (also found in a vegetable diet) [39]. Botanical Drugs pharmacology requires a knowledge about the actual ligand concentration at a given receptor site. This is fundamental to understand the pharmacodynamic behavior of one compound in a physiological context (i.e., ligand + receptor associated vs. free states) and ultimately the mechanism of action. If in a mixture of several hundred potentially bioactive compounds, one should know what the plasma and tissue concentrations of the individual compounds are and all of the respective receptor interactions (Kd values) at the given concentration in order to understand how it works. Both maximal concentrations and half times would provide the basis for the feasibility of postulated mechanisms of action from in vitro studies [34].

These above Pharmacokinetics and Pharmacodynamics challenges can be solved with Translational Research Model in Botanical Drugs. Translational Medicine approach can be very well applied to novel Botanical Drug Product Development. The translational research objective is to reduce time and optimize input to translate research to practice and further then its impact on healthcare and distinguish its value addition throughout. Thus, one can improve quality of research in same resources and optimize cost in perform-ing research project and then weigh it against anticipated reward.

The Integrated outline of Translational Research Model is explained in Figure 4.3.

Translation Research Model in Botanical Drug Development can have the following parameters assessed throughout its process marker development as Success Measure Scale and offer the logical explanation in sequence (Figure 4.4).

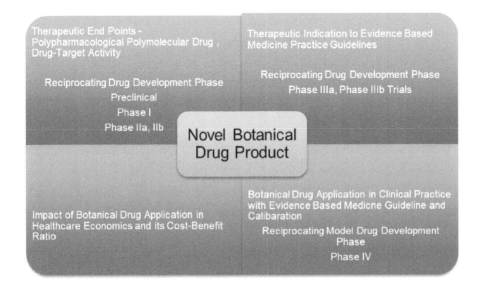

Figure 4.3 *Integrated model for Botanical Drug Product with translational research.*

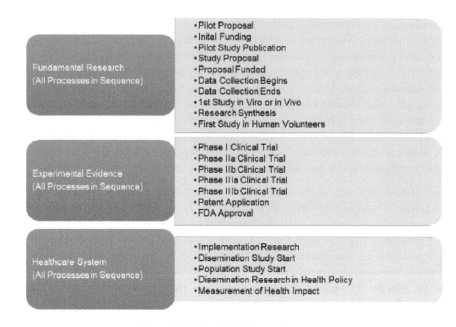

Fundamental Research
(All Processes in Sequence)
- Pilot Proposal
- Inital Funding
- Pilot Study Publication
- Study Proposal
- Proposal Funded
- Data Collection Begins
- Data Collection Ends
- 1st Study in Viro or in Vivo
- Research Synthesis
- First Study in Human Volunteers

Experimental Evidence
(All Processes in Sequence)
- Phase I Clinical Trial
- Phase IIa Clinical Trial
- Phase IIb Clinical Trial
- Phase IIIa Clinical Trial
- Phase IIIb Clinical Trial
- Patent Application
- FDA Approval

Healthcare System
(All Processes in Sequence)
- Implementation Research
- Disemination Study Start
- Population Study Start
- Disemination Research in Health Policy
- Measurement of Health Impact

Phase II Clinical Trial Integration

Proposal Submission → IRB Approval → Proposal Funded → Trial Start to End → Results Presented & Published

IRB Approval

IRB Proposal → IRB Approval Review → IRB Proposal Revision → IRB Proposal Resubmit → IRB Commitee Review → Proposal to Revise → Proposal Resubmit and Approval

Figure 4.4 *Process markers in translation research. (Adapted from Trochim, W. et al., Clin. Transl. Sci., 4, 153–162, 2011 [40].)*

- Therapeutic Indication to target with possible available interventions with SWOT Analysis
- Target in Therapeutic Intervention
- Novel Drug Molecule Structure – Polypharmacological in nature
- Novel Drug Molecule Function – Target Validation (Theoretical)
- Validation Consistency through reproducible experiments
- Degree of Logic in Drug to Target to its Healthcare Impact
- Specificity in Novel Drug Possibility – Therapeutic Intervention-its Practical Impact-Benefit Logic
- Validity and Outcome of Logic through Clinical Trials Experiment Evidence
- Health Care Impact to Cost – Benefit Ratio [41–43]

4.4 Role of Ayurveda, Indian System of Medicine in developing Botanical Drug Products?

The World Health Organization's Commission on Intellectual Property and Innovation in Public Health has recognized the promise and role of traditional medicine like Indian System of Medicine in drug development for affordable health solutions [44].

Ayurved (Indian System of Medicine) is one of the most ancient medical sciences known to human kind. "Ayurved" is Sanskrit Word that literally means "Science of Life." As per Modern Astronomical Calculations it was originated way back 24,000 years. It was taught in Indian Medical Schools through "Oral Tradition," however its written medical scriptures, Principles and Practices, are at least 5,000 years old. Ayurved is sub part of "Atharvaveda," which is one of the four most important Knowledge Series known to human kind. Ayurved is subjected to not only Human Beings but also to Veterinary, Environment, Agriculture, and Architecture practices. Its basic principles are applied in Physics, Mathematics, Chemistry, and Mechanics as well. Ayurved has eight Medical Branches—Internal Medicine, Pediatrics, ENT, Surgery, Toxicology and Medical Jurisprudence, Psychiatry, Geriatrics and Disease Reversal Medicine, and Progeny Medicine.

Combining the strength of knowledge based on traditional systems such as Ayurveda (Indian System of Medicine) with the dramatic power of combinatorial sciences and high-throughput screening (HTS) will help in the generation of structure activity libraries. The three main hurdles in the drug development to provide new functional leads are time, money, and toxicity that can be reduced by Ayurvedic Knowledge and Experimental Database [45].

- *Time*: Through Indian System of Medicine one can find at least 10,000 novel leads for different kind of therapeutic interventions through Observational Therapeutics. Such leads can save at least 4–5 years of time in entire drug discovery and development process.
- *Money*: If a qualified drug candidate fails in therapeutic efficacy in later stage of drug development, the opportunity cost is very high; thus, prequalified therapeutic indication through thousands of years Indian System of Medicines therapeutic observation can cut down failure risk significantly.
- *Toxicity*: As Indian System of Medicine is at least 10,000 years old, most of the Poly Combinations of Botanicals are being used so far has real-time safety profile; thus, one may not require toxicity studies to be conducted and resources can be saved.

With the information and knowledge of nearly 35,000 classical preparations of Ayurveda, Indian System of Medicine for various therapeutic indications one can discover successful leads for Complex Disorders of modern age. The Drug Discovery Process Pathway to develop a Botanical Drug Products can be enumerated as below (Figure 4.5 and Table 4.1).

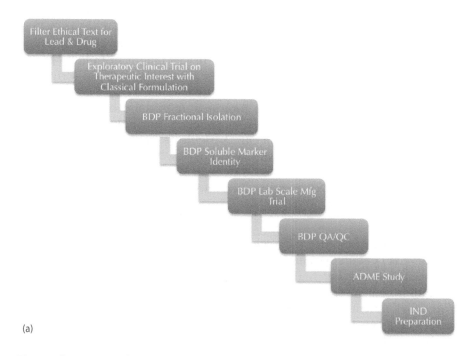

(a)

Figure 4.5 *Strategy for Botanical Drug Product through Ayurveda, Indian System of Medicine.* (*Continued*)

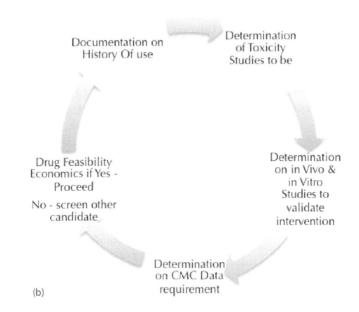

Figure 4.5 (Continued) Strategy for Botanical Drug Product through Ayurveda, Indian System of Medicine.

Table 4.1 Chemistry, Manufacturing, Control Data Required for Botanical Drug Product of Ayurveda, Indian System of Medicine

Botanical raw material	• Identification by trained personnel — Pharmacognosy • Certificate of authenticity — BRM standardization • List of all growers and/or suppliers — uniformity in supply chain
Botanical drug substance	• Qualitative and quantitative description — soluble marker analysis • Name and address of manufacturer • Description of manufacturing process — uniformity in process control • Quality control tests performed • Description of container/closure system • Available stability data • Container label
Botanical drug product	• Qualitative description of finished product composition of finished product — with drug delivery system • Name and address of manufacturer • Description of manufacturing process • List of quality control tests performed • Description of container/closure system • Available stability data • Placebo • Labeling • Environmental assessment or claim of categorical exclusion

Source: Adapted from CMC Requirements for Herbal Products POS.pdf. Hurley Consulting Associates [46].

Table 4.2 Fundamental Principles of Indian System of Medicine and Its Important Application in Botanical Drug Product Discovery and Development

No.	Principle	Important Applications in Botanical Drug Discovery and Development Pathway
1.	Purusha-Prakriti Siddhanta	Identification of appropriate target in selected disease
2.	Sapta-Padartha Siddhanta	Characterization of Disease Target and Ligand & Safety & Efficacy Determination of Ligand
3.	Dosha-Dhatu-Mala Siddhanta	Bioavailability and Bioequivalence Determination of Ligand
4.	Purusha-Loka Siddhanta	Ligand discovery

Table 4.3 Other Valuable Concepts from Indian System of Medicine Can Be Applied in Botanical Drug Product Discovery and Development

No	Principle	Important Applications in Botanical Drug Discovery and Development Pathway
1.	Srotasa and Maheshwara Sutra	Target – Ligand Validation and Mode of Action
2.	Marma Sharir	Transdermal Drug Delivery System
3.	Agni Samskar & Dhatu Saratva	ADME and Bio-adaptability of Ligand
4.	Shat-Kriya Kala and Roga Marga	Clinical Trial Design, Dosage and Therapeutic Primary and Secondary End Points Determination
5.	Samhitas	Ligand development through suitable Botanical Raw Material and Drug Substance
6	Bhaishajya Kalpana & Ras Shastra	Chemistry, Manufacturing and Control and Botanical Drug Product design and manufacturing process

There are valuable Basic Principles of Ayurveda, Indian System of Medicine that can be applied in discovering novel Botanical Drug Product in Complex Disorder Intervention (Tables 4.2 and 4.3).

4.4.1 Purusha-Prakriti Siddhanta

Ayurveda has offered conspicuous theory of Origin of Universe Manifestation through Sankhya Darshana. Purusha-Prakriti Siddhanta constitutes 24 principles and in permutation and combinations thereof one can understand very nature, constitution, behavior of every single living and non-living substances at sub atomic, quantum level. These 24 principles are:

Purusha, Prakriti (Triguna), Mahat "Buddhi," Ahamkara, Manas (5 in number)
Tanmatra – Shabda, Sparsha, Rupa, Rasa, Gandha (5 in number)
Panchabhuta – Pruthvi, Aap, Teja, Vayu, Akasha (5 in number)
Pancha Gyanendriya – Shrotra, Twacha, Netra, Jivvha, Ghrana (5 in number)
Pancha Karmendriya – Pada, Hasta, Jivvha, Upastha, Guda (5 in number)

Any disease condition can be analyzed with the help of these principles and one can define appropriate target for that disease condition.

4.4.2 Sapta-Padartha Siddhanta

With this fundamental principle, Ayurved can put forward the micro analysis of objects and its functioning.

Dravya (9 in number) –
Pruthvi, Jala, Teja, Jala, Vayu, Aksha, Mana, Kaal, Disha, Atma
Guna – four types

- Gurvadi – Guru, Sheeta, Snigdha, Manda, Sthira, Mrudu, Vishada, Shlashna, Sukshma, Sandra, Lagu, Ushna, Ruksha, Tikshna, Sara, Kathina, Pishhila, Khara, Sthula, Drava (20 in number)
- Paradi – Para, Apara, Yukti, Sankhya, Sanyoga, Vibhag, Pruthakatva, Pariman, Sanskar, Abhyasa (10 in number)
- Visheha – Shabda, Sparsha, Rupa, Rasa, Gandha (5 in number)
- Adhyatmika – Ichchha, Dwesha, Sukha, Dukhha, Prayatna, Buddhi (6 in number)

Karma – Utkshepana, Apakshepana, Akunchana, Prasarana, Gamana (5 in number)
Samanya – Dravya Samanya, Guna Samanya, Karma Vishesha (3 in number)
Vishesha – Dravya Vishesha, Guna Vishesha, Karma Vishesha (3 in number)
Samavaya – Samavayi, Asamavayi, Nimitta (3 in number)
Abhava – Prak Abhava, Pradhwansa Abhava, Atyanta Abhava, Anyonya Abhava (4 in number)

Considering these 64 Principles together Safety and Efficacy Range of Ligand can be determined and the impact of Time and Space on Matter and Consciousness can be understood. This knowledge can be helpful in understanding characterization of disease target as well as ligand dynamics in real time and space reference.

4.4.3 Dosha-Dhatu-Mala Siddhanta

Ayurved has established metabolic transformation through three sets of principles

Dosha – (3 in number)
Vayu – Prana, Vyana, Samana, Udana, Apana (5 sub types)
Pitta – Sadhaka, Bhrajaka, Alochaka, Pachaka, Ranjaka (5 sub types)
Kapha – Tarpaka, Avalambaka, Kledaka, Bodhaka, Shleshmaka (5 sub types)
Dhatu – Rasa, Rakta, Mansa, Meda, Asthi, Majja, Shukra (7 in number)
Mala – Purisha, Mutra, Sveda (3 in number)

Assesing Panchamahabhuta Constitution of Ligand with Panchamahabhuta constitution of Dosha-Dhatu-Mala can yield the idea of Pharmacokinetics and Pharmacodynamics range of Ligand.

4.4.4 Purusha Loka Siddhanta

Ayurved extends clear relationship nature in between Micro Cosmos and Macro Cosmos and vice versa through 64 Principles as explained in Sapta Padartha Siddhanta. Through this principle many Ligands can be discovered through Observational Therapeutics.

4.4.5 Srotasa and Maheshwara Sutra

This is one of the distinguished ideas of Ayurved where in it explains how various physiological systems operates at Systemic, Tissue, Molecular, Atomic, Sub Atomic, and Quantum levels in terms of Systems Biology.

Srotasa (14 in number) – Prana, Anna, Udaka, Rasa, Rakta, Mansa, Meda, Asthi, Maja, Shukra, Artava, Purisha, Mutra, Sweda

Maheshwara Sutra (14 in number) – this view comprises the exact mechanism of origin and manifestation of every object and its frequency in terms of strings.

अ इ उ ण्
ऋ लृ क्
ए ओ ङ्
ऐ औ च्
ह य व र ट्
ल ण्
ञ म ङ ण न म्
झ भ ञ्
घ ढ ध ष्
ज ब ग ड द श्
ख फ छ ठ थ च ट त व्
क प य्
श ष स र्
ह ल

The knowledge of Target – Ligand Validation and Mode of Possible Therapeutic Action as in Network Pharmacology can be derived through application of these principles.

4.4.6 Marma Sharir (108 in number)

As per Ayurved there are 108 potentially sensitive anatomical and physiological areas on human body that are important in terms of Transdermal Drug Delivery Systems. Botanical Drug Products can be delivered at respective anatomical sites through dermal and or injectable routes very efficiently to gain desired plasma levels of drug.

4.4.7 Agni Sanskara and Dhatu Saratva

Agni (13 in number) – Rasagni, Raktagni, Mansagni, Medagni, Asthiagni, Majjagni, Shukragni, Koshthagni, Pruthvi agni, Appagni, Tejagni, Vayvyagni, Vyomagni

Dhatu Sara (8 in number) – Rasa, Rakta, Mansa, Meda, Asthi, Majja, Shukra, Oja

"Agni" is one of the profound concepts of Ayurved that simply denote metabolic transformation at cellular level. "Dhatu Saratva" means which Physiological System is comparatively more efficient metabolically than others. Combination of these two concepts can bear determination on Absorption, Distribution, Metabolism, and Excretion of Botanical Drug along and its therapeutic site of action.

4.4.8 Shat Kriya Kaala and Roga Marga

Shat Kriya Kala (6 in number) – Sanchaya, Prakopa, Prasar, Sthana Sanshraya, Vyakti, Bheda

Roga Marga (3 in number) – Koshtha, Shakha, Marmasthi

This is a unique outlook of Ayurveda in terms of defining Etiology and Pathogenesis of any Disease so that the knowledge of potential therapeutic intervention points can be conceptualized. This outlook also can be useful in determining Clinical Trial Design for novel Botanical Drug Product and Primary and Secondary End Points.

4.4.9 Samhita medicinal literature of Ayurveda (6 in number)

Bruhat Samhita Trayi– Charaka, Sushruta, Vagbhatta (3 in number)

Laghu Samhita Trayi– Sharangdhar, Bhava Prakasha, Madhava Nidana (3 in number)

These basic medicinal literatures of Ayurved with at least 20 different Nighantu (Medicnal Plants Compendium) have definite novel Botanical Drug Products leads for various Complex Disorders. These literatures also provide significant insights about creating Botanical Drug Product Polymorphs for same disease so that different versions of Botanical Drugs can be process fabricated with the help of Pharmacogenomics.

4.4.10 Rasa Shastra and Bhaishajya Kalpana

These branches of Ayurved predominantly deal with Pharmaceutical Engineering. There are various concepts and tools established through which Botanical Drug Products can be successfully designed. Also the detailed description of how various metals and minerals can be incorporated with Botanical Drugs and enhance their therapeutic efficacy. Sometimes different processes yield different molecular nature or soluble markers in Botanical Drug Products so this knowledge can cut down the risk of CMC validation and streamlining regulatory approval of process and sites.

4.5 Traditional Knowledge Digital Library

To organize Indian Traditional Knowledge and protect it from Intellectual Property exploitation, Government of India has undertaken an ambitious project of creating a Traditional Knowledge Digital Library (TKDL). This is a joint venture of the Council of Scientific Research and Central Council for Research in Ayurveda and Siddha. This project is intended to cover about 35,000 formulations available in 14 classical texts of Ayurveda to convert the information in to patent compatible format. The work has been initiated with a co-operative set up of 30 Ayurveda experts, 5 Information Technology experts, and 2 Patent examiners. The digital library will include all details in digital format about international patent classification, traditional research classification, Ayurveda terminology, concepts, definitions, classical formulations, doses, disease conditions and references to documents [47].

4.6 Conclusion

Reverse Pharmacology Techniques rather than Forward Pharmacology Techniques can be certainly useful in developing new drugs for Complex Disorders. In conjugation with knowledge of centuries old and practiced and tested ancient medicinal sciences like Ayurved can be of utmost importance as it can provide a semi-qualified drug lead hypothesis and reduce preliminary drug development cost significantly along with certain probability of clinical success. Translational Research Model to conduct Clinical Trials especially in the case of Botanical Drug Products is more robust as new drug interventions outcome on healthcare quality can be mapped.

References

1. 2007 FDA drug approvals: A year of flux. Available at https://www.nature.com/articles/nrd2514.
2. B. Chen, D. Wild and R. Guha, PubChem as a source of polypharmacology, *J. Chem. Inf. Model.* 49 (2009), pp. 2044–2055.

3. S. Frantz, Pharma faces major challenges after a year of failures and heated battles, *Nat. Rev. Drug Discov.* 6 (2007), pp. 5–7.

4. A.D.B. Vaidya, Reverse pharmacological correlates of ayurvedic drug actions, *Indian J. Pharmacol.* 38 (2006), p. 311.

5. R. Gupta, B. Gabrielsen and S.M. Ferguson, Nature's medicines: Traditional knowledge and intellectual property management: Case studies from the national institutes of health (NIH), USA, *Curr. Drug Discov. Technol.* 2 (2005), pp. 203–219.

6. K.B. Seamon, W. Padgett and J.W. Daly, Forskolin: Unique diterpene activator of adenylate cyclase in membranes and in intact cells, *Proc. Natl. Acad. Sci. USA.* 78 (1981), pp. 3363–3367.

7. B. Das, V. Tandon, L.M. Lyndem, A.I. Gray and V.A. Ferro, Phytochemicals from Flemingia vestita (Fabaceae) and Stephania glabra (Menispermeaceae) alter cGMP concentration in the cestode Raillietina echinobothrida, *Comp. Biochem. Physiol. Toxicol. Pharmacol. CBP* 149 (2009), pp. 397–403.

8. M. Khan, K. Vimal, A. Gajjar, N. Vyas, P. Siddharth and B. Amee, Reverse pharmacognosy in new drug discovery, *J. Curr. Pharm. Res.* 1 (2007).

9. F. Ntie-Kang, D. Zofou, S.B. Babiaka, R. Meudom, M. Scharfe, L.L. Lifongo et al., AfroDb: A select highly potent and diverse natural product library from African medicinal plants, *PLOS ONE* 8 (2013), pp. e78085.

10. Q.-T. Do and P. Bernard, Pharmacognosy and reverse pharmacognosy: A new concept for accelerating natural drug discovery, *IDrugs Investig. Drugs J.* 7 (2004), pp. 1017–1027.

11. C.N. Cavasotto, Homology models in docking and high-throughput docking, *Curr. Top. Med. Chem.* 11 (2011), pp. 1528–1534.

12. P. Ripphausen, B. Nisius, L. Peltason and J. Bajorath, Quo vadis, virtual screening? A comprehensive survey of prospective applications, *J. Med. Chem.* 53 (2010), pp. 8461–8467.

13. T. Cheng, Q. Li, Z. Zhou, Y. Wang and S.H. Bryant, Structure-based virtual screening for drug discovery: A problem-centric review, *AAPS J.* 14 (2012), pp. 133–141.

14. S. Okamoto, M.A. Pouladi, M. Talantova, D. Yao, P. Xia, D.E. Ehrnhoefer et al., Balance between synaptic versus extrasynaptic NMDA receptor activity influences inclusions and neurotoxicity of mutant huntingtin, *Nat. Med.* 15 (2009), pp. 1407–1413.

15. G. Bottegoni, I. Kufareva, M. Totrov and R. Abagyan, Four-dimensional docking: A fast and accurate account of discrete receptor flexibility in ligand docking, *J. Med. Chem.* 52 (2009), pp. 397–406.

16. D.B. Kokh and W. Wenzel, Flexible side chain models improve enrichment rates in in silico screening, *J. Med. Chem.* 51 (2008), pp. 5919–5931.

17. H.J. Böhm, The computer program LUDI: A new method for the de novo design of enzyme inhibitors, *J. Comput. Aided Mol. Des.* 6 (1992), pp. 61–78.

18. G. Jones, P. Willett, R.C. Glen, A.R. Leach and R. Taylor, Development and validation of a genetic algorithm for flexible docking, *J. Mol. Biol.* 267 (1997), pp. 727–748.

19. B.K. Shoichet and I.D. Kuntz, Matching chemistry and shape in molecular docking, *Protein Eng.* 6 (1993), pp. 723–732.

20. M.A. Lie, R. Thomsen, C.N.S. Pedersen, B. Schiøtt and M.H. Christensen, Molecular docking with ligand attached water molecules, *J. Chem. Inf. Model.* 51 (2011), pp. 909–917.

21. S.-Y. Huang and X. Zou, Inclusion of solvation and entropy in the knowledge-based scoring function for protein-ligand interactions, *J. Chem. Inf. Model.* 50 (2010), pp. 262–273.

22. R. Abel, T. Young, R. Farid, B.J. Berne and R.A. Friesner, Role of the active-site solvent in the thermodynamics of factor Xa ligand binding, *J. Am. Chem. Soc.* 130 (2008), pp. 2817–2831.

23. L. Wang, B.J. Berne and R.A. Friesner, Ligand binding to protein-binding pockets with wet and dry regions, *Proc. Natl. Acad. Sci. USA.* 108 (2011), pp. 1326–1330.

24. D. Hecht and G.B. Fogel, Computational intelligence methods for docking scores, *Curr. Comput. Aided Drug Des.* 5 (2009), pp. 56–68.

25. Z. Deng, C. Chuaqui and J. Singh, Structural interaction fingerprint (SIFt): A novel method for analyzing three-dimensional protein-ligand binding interactions, *J. Med. Chem.* 47 (2004), pp. 337–344.

26. New Approaches to Virtual Screening. Available at https://www.dddmag.com/article/2013/12/new-approaches-virtual-screening.

27. P. Chène, Can biochemistry drive drug discovery beyond simple potency measurements? *Drug Discov. Today* 17 (2012), pp. 388–395.

28. A. Cortés-Cabrera, F. Gago and A. Morreale, A reverse combination of structure-based and ligand-based strategies for virtual screening, *J. Comput. Aided Mol. Des.* 26 (2012), pp. 319–327.

29. P. Yeh and R. Kishony, Networks from drug-drug surfaces, *Mol. Syst. Biol.* 3 (2007), p. 85.

30. J.B. Fitzgerald, B. Schoeberl, U.B. Nielsen and P.K. Sorger, Systems biology and combination therapy in the quest for clinical efficacy, *Nat. Chem. Biol.* 2 (2006), pp. 458–466.

31. Reverse pharmacology and systems approaches for drug discovery and development. Available at http://www.eurekaselect.com/68105/article.

32. P.N. Leão, A.R. Pereira, W.-T. Liu, J. Ng, P.A. Pevzner, P.C. Dorrestein et al., Synergistic allelochemicals from a freshwater cyanobacterium, *Proc. Natl. Acad. Sci. USA.* 107 (2010), pp. 11183–11188.

33. K. Shiojiri, R. Ozawa, S. Kugimiya, M. Uefune, M. van Wijk, M.W. Sabelis et al., Herbivore-specific, density-dependent induction of plant volatiles: Honest or "cry wolf" signals? *PLOS ONE* 5 (2010), pp. e12161.

34. J. Gertsch, Botanical drugs, synergy, and network pharmacology: Forth and back to intelligent mixtures, *Planta Med.* 77 (2011), pp. 1086–1098.

35. M.A. Fischbach and J. Clardy, One pathway, many products, *Nat. Chem. Biol.* 3 (2007), pp. 353–355.

36. M.A. Fischbach, C.T. Walsh and J. Clardy, The evolution of gene collectives: How natural selection drives chemical innovation, *Proc. Natl. Acad. Sci. USA.* 105 (2008), pp. 4601–4608.

37. A.L. Hopkins, Network pharmacology: The next paradigm in drug discovery, *Nat. Chem. Biol.* 4 (2008), pp. 682–690.

38. J. Gertsch, J.M. Viveros-Paredes and P. Taylor, Plant immunostimulants—scientific paradigm or myth? *J. Ethnopharmacol.* 136 (2011), pp. 385–391.

39. S. McLean and A.J. Duncan, Pharmacological perspectives on the detoxification of plant secondary metabolites: Implications for ingestive behavior of herbivores, *J. Chem. Ecol.* 32 (2006), pp. 1213–1228.

40. W. Trochim, C. Kane, M.J. Graham and H.A. Pincus, Evaluating translational research: A process marker model, *Clin. Transl. Sci.* 4 (2011), pp. 153–162.

41. R. Doll, Sir Austin Bradford Hill and the progress of medical science, *BMJ* 305 (1992), pp. 1521–1526.

42. R. Doll, Sir Austin Bradford Hill, 1897–1991, *Stat. Med.* 12, pp. 795–808.

43. A.B. Hill, The environment and disease: Association or causation? *Proc. R. Soc. Med.* 58 (1965), pp. 295–300.

44. B. Patwardhan, *Traditional Medicine: Modern Approach for Affordable Global Health*, WHO, Geneva, Switzerland, 2005.

45. B. Patwardhan and M. Hooper, Ayurveda and future drug development, *Int J Altern. Complement Med* 10 (1992), pp. 9–11.

46. CMC Requirements for Herbal Products POS.pdf. Hurley Consulting Associates.

47. TKDLTraditional Knowledge Digital Library. Available at http://www.tkdl.res.in/tkdl/LangDefault/Common/Home.asp?GL=Eng.

5

Industrial Scale-Up Strategy and Manufacturing Technologies for Botanical Drug Products

Rahul Maheshwari, Kaushik Kuche, Ashika Advankar, Namrata Soni, Piyoosh Sharma, Muktika Tekade, and Rakesh Kumar Tekade

Contents

Disclosures: There is no conflict of interest and disclosures associated with the manuscript.

€ = *Authors with equal contribution and can be interchangeably written as first author*

5.1 Introduction to botanical drug products

In recent time, products obtained from plants to be used as bioactive substance are of prime interest. The reason behind this belongs to an ever increasing demand of products based on herbal concept (Ansari, 2016). The constituents derived from plants for medicinal use can be derived from one or more parts such as, root, leaf, stem, algae, fungi, lichens, etc. These plant parts as raw material then go through further manufacturing processes like extraction, distillation, purification, concentration, fermentation, and further down stream processing (Mariod, 2016). The botanical products sometimes can be labelled as natural foods or sport supplements in various countries. These products can be treated as therapeutic goods and food additives based on their application (Villarreal, 2017).

This current time belongs to the development of products with desired characteristics by making the use of nanotechnology in medicines (Lalu et al., 2017, Tekade et al., 2017b). Moreover, the combination of nanotechnology and botanical drug products development may result into better therapeutic effectiveness (Maheshwari et al., 2015b, Soni et al., 2016). Many variants of drug delivery like liposomes (Maheshwari et al., 2012, 2015a), dendrimers (Soni et al., 2017), and polymeric nanoparticles (Kumar Tekade et al., 2015, Sharma et al., 2015) may be utilized in the development of botanical drug products.

Plant-based products are generally accessible to buyers through various distribution channels. Specifically, they are sold over the counter in drug stores and can be purchased additionally in general stores, botanist's shops, or through the Internet (Sengar et al., 2017). Employment of plant products to cure or to treat human diseases has grown worldwide, depending on local environment, local medical practices, and nutritional practices, availability of plant species of interest, and main protocols of customary industries on the territory (Van Wyk and Wink, 2017).

In context, there are several ways in which individual countries are defining botanical products or herbs, or herbal drugs derived from them, and countries have implemented numerous methods to license, dispense, manufacture, and trade to ensure their efficacy, safety, and quality. Because of these reasons botanical products differ from country to country. In addition, national or local bodies of individual country has also enabled and addressed either by direct way or indirect way the marketing of conventional botanical products as health supplements or as therapeutic products (Brown, 2016). Nanocarrier approaches may avoid many formulation related hurdles such as solubility, stability, and desired pattern of release and may help in developing botanical drug products (Tekade et al., 2017a, 2017c, 2017d).

5.2 Chemotaxonomy-different chemical groups in medicinal plants

In general, the classification of plants is based on their morphology, anatomy, and chemotaxonomy. Out of these systems of classification, the first two are traditional, while the third—chemotaxonomy—is relatively a newer method. The word chemotaxonomy is the combination of "chemo" and "taxonomy," which means "the classification system of plants on the ground of their chemical composition" (Zhu et al., 2016). Before the existence of chemotaxonomy the classification of plant species involved the use of some factors relating to morphology/anatomy such as edibility, taste, odor, and color, without having considerable info about chemical nature (Julier et al., 2016).

It is exciting to know that there is increasing interest of scientists toward chemotaxonomy because of the peculiar nature and structure of the basic chemical moiety of secondary metabolites. In taxonomy, secondary metabolites played an important role due to their specificity and therefore serve as an effective tool for classification. However, for precise characterization of plant species, sound knowledge of the biosynthetic pathways of these metabolites is required. This amalgamation of examining chemical composition of targeted plant with the morphological analysis leads to edge cutting tool for their characterization and classification (Silva et al., 2017).

Moreover, reports are available supporting that the phytochemical abundance, genetic data and morpho-anatomic characteristics are still sparingly exist for some genera and therefore the addition of chemotaxonomic perspective to available database will give a more significant phylogenetic distribution, because of the presence of secondary metabolites which are naturally produced in taxon (Srivastava et al., 2017). Furthermore, as the primary metabolites are ubiquitously distributed, they are not appropriate target for chemotaxonomic distribution of plants. Therefore, the distribution profile of secondary metabolites in plants brought additional knowledge for the taxonomic classification. The approaches employed for chemotaxonomic distribution is more significant and modern than traditional one and supports advanced investigations of current time of botanical product development, either it is procurement or processing of plant species or quantitative and qualitative analysis of metabolic products (Bakoğlu et al., 2017).

The advancement of analytical chemistry with modern technologies, such as high-performance liquid chromatography (HPLC), ultra-performance liquid chromatography (UPLC) over pressure layer chromatography (OPLC), Fourier transform infrared spectroscopy (FT-IR), and liquid chromatography etc. in qualitative and quantitative analysis of chemical profile of targeted compounds up to trace amount of µg and/or ng leads to rise and strengthening of chemotaxonomy as a branch of plant classification (Sharma et al., 2017).

Very recently, Julier et al. (2016) demonstrated the successful use of FT-IR to interpret the cryptic diversity of Poaceae pollen (powder like substance composed of pollen grains which are male microgametophytes of seed plants, which produce male gametes (sperm cells)). Authors showed that investigation of FTIR spectra from a selection of Poaceae taxa could be employed to characterize pollen grains. Calculations based on nearest neighbor classification algorithm showed similar results that were previously obtained using very-expensive and time-consuming approach viz scanning electron microscopy (SEM).

The chemotaxonomy technique of plant classification has become evident from some currently reported investigations considering some potential plant species in several genera such as Malvaceae, Ranunculaceae, Magnoliaceae, Polygonaceae, and Solanaceae (Rosuman and Lirio). Not limited to elucidate chemical classification of plant species, chemotaxonomic investigations could be extended further to benefit humans in exploration of clinically effective and medicinal plants. Several studies have been reported for chemotaxonomic characterization of different genera and families for better understanding of potential species viz. Cannabis, Acacia, Withania, Ornithine-derivative alkaloids, and sometime also as a tool for authentication of herbal drug i.e., Chamomile etc (Misra and Srivastava, 2016).

5.3 Different chemical properties of phytochemicals

Phytochemicals are made up of many chemical compounds having defined structures. These mixtures of different compounds are responsible for many pharmacological activities. Chemically phytochemicals are very diverse in nature, having divided into numerous classes according to their chemical characteristics (Sánchez-Salcedo et al., 2015). These structures range from very small and simple structures to big and complex ones. These chemical compounds may be responsible for the color of the compound, e.g., anthocyanins are water-soluble-colored pigment present in various plants. These activities and properties may be attributed to the presence of functional group. The functional groups are an important factor in deciding which chemicals reactions a compound may undergo (Ahmed et al., 2014). Various reactions are present, which may help in identify a compound based on function group present in it. Table 5.1 represents functional groups and the tests used for their identification.

There are also various tests on the basis of chemical class of a compound such as for alkaloids Hager's test, Wagner's test, Frohde's test and many other chemical reactions can be done. For flavonoids we have Shinoda test; for steroids we have Liebermann Burchard and Salwoski's test; and for sugar we have Molisch test (Yadav et al., 2017).

Polarity of compounds present may also play an important role. Hence, to isolate polar compounds from a crude drug, polar solvents such as dichloromethane, methanol, and water are used. In 2017, isolated 28 polar compounds by isolated

Table 5.1 Depicting Various Functional Groups Present in Phytochemicals and the Reaction for Their Identification

Functional Group	Reaction for Its Identification in a Compound
Hydroxyl	a. Alkylation
	b. Benzoylation
Phenolic group	a. With $FeCl_3$ gives blue or green colour
	b. Addition of NaOH dissolves the compound and is reprecipitated using CO_2
Ketone	a. With hydroxylamine gives corresponding oxime
	b. With 2,4-dinitropyridine gives corresponding orange coloured phenylhydrazone
	c. With semicarbazine gives semicarbazone
Aldehyde	a. Positive silver mirror test
Acid	a. Solubility in sodium bicarbonate
	b. Solubility in ammonia
	c. Reprecipitated with CO_2
Ester	a. With hydroxyl amine and ferric chloride gives magenta dye
	b. On base hydrolysis with NaOH gives alcohol and sodium salt

them in chloroform-methanol and methanol extract (Camero et al., 2017). Alabri et al. (2014) prepared extracts of *Datura metel* in different solvents from polar to non-polar and performed preliminary phytochemical screening on all those extracts to know the class of compound. Methanolic extract which was polar in nature did not show presence of steroids and triterpenoid which are non-polar compounds but showed positive test for glycosides and flavonoids. Similarly, in extracts in polar solvents showed presence of polar compounds while non-polar solvent showed presence of non-polar compounds.

5.4 Factors effecting determination of isolation strategy

There are several factors that eventually determines the quality and level of extraction, out of which choice of solvent (menstruum) is the most critical also the nature of the drug used (botanical structure) and solubility of constituents and particle size of the feed.

5.4.1 Solvent for extraction

Usually the solvent selected for extraction must be selective for the compound to be extracted, should possess a high capacity for extraction, must be available at low costs and easy to remove from the obtained extract, and most importantly it must not react with the constituents which it is supposed to isolate. Broadly used solvents for isolation are water, ether, alcohol, chloroform etc. Vital factors that govern the selection of solvent are the quantity of phytochemical to be extracted, rate at which material is to be extracted, diversity of compounds to be extracted, toxicity associated with solvent in bioassay process and the potential health dangers associated with extracts (Chandrasekara et al., 2016).

Water has several benefits of being used as a solvent for extraction but majority of the active compounds that are to be isolated from plants are water insoluble (Wilson et al., 2014). Certain constituents, like starches, gums, sugars, can be extracted using water but not acceptable as it effects the clarity of the preparation, further water also serves to be an excellent growth promoting media for bacteria, yeast and moulds thus, preservatives are to be added which further increase the manufacturing cost. Alcohols could be used instead as it can solubilize several constituents, thus are employed for extraction of alkaloids, vitamin oils, glycosides, and resins also, it can dissolve coloring constituents like tannins and other organic acids and salts (Azmir et al., 2013).

5.4.2 Particle size of the feed

The feed or the herb is generally subjected to size reduction that leads to achieve optimum extraction surface area that also depends on the botanical structure of the drug like, drugs are sliced for drugs like gentian, coarse to moderately coarse in case of cascara and belladonna, moderately fine for woody drugs and coarse powder in case if the drug is leafy. Ideally it is said that if the size of the drug is reduced to individual cell, then the extraction and isolation could be done at its highest efficiency; however, such cellular-level-size reduction is not possible and not recommended as it could lead to decomposition of active constituents and if constituent is volatile then it would be lost (Avio et al., 2015).

Further are cellular level particles being diluted chances arises that cell may get burst and empty their all constituents in the solvent thereby making the isolation process even more tough. Thus, it is recommended that size reduction should be optimum so that there should be increased surface area by distortion or breakage of cells that facilitates solvent penetration and escape (with the soluble content) and decreased radial distance that helps in maintaining the concentration gradient (Allen, 2013).

5.4.3 Solubility factors

In the extraction process, not only the solvent and size but exchange of constituents is also a critical point. This exchange of constituents is depended on the solubility of constituent in the solvent thus, governed by the Noyes Whitney equation given as follows,

$$\frac{dC}{dt} = \frac{DA(Cs - Cb)}{h}$$

where: dC/dt = rate of dissolution; D = diffusion coefficient; A = surface area of the herb/feed; Cs = concentration of solute particle at boundary layer; Cb = concentration of solute in bulk.

The dissolved material (desired constituent) must reach the surface of the particle and should pass through boundary layer at interface, wherein the rate of diffusion depends on concentration gradient between the center of particle, outwards and boundry layer (Tomita et al., 2014). The diffusion rate also depends on the thickness of the boundary layer and the diffusion coefficient of the solute. Thus, higher the dissolution and diffusion, the higher would be the extraction rate, hence all the mentioned factors in the noyes whitney equation play a vital role in the extraction process. Other factors that can affect the extraction are as follows.

5.4.3.1 When drug immersed in solvent

When the drug is completely immersed in solvent, and if it is stirred continuously, then the concentration gradient by dispersing the local concentration and thus reducing the thickness of boundary layer hence, enhancing the extraction process. The concentration gradient against the boundary layer can also be increased by suspending the drug in the cloth or over a perforated plate just near the surface of liquid, wherein the drugs dissolve and density of the solution increases leading to the convection of solution thus serving the purpose of increased concentration gradient (Chen et al., 2015).

5.4.3.2 When solvent is passed across the drug particles

As the solvent is flowed, the saturated solution is replaced with fresh solvent rapidly also the flow of solvent reduces the boundary layer thereby helping in maintaining the concentration gradient, this can be achieved by placing the sample in perforated sample (Azmir et al., 2013).

5.4.3.3 Elevated temperature effect

As the temperature of the chemical reaction increases, the extraction efficiency of solvent increases due to the solubility of majority of constituents increases in the solvent. Thus it increases extraction efficiency and product yield (Meglič et al., 2016).

5.5 Global nutraceutical market for botanical drug products

As per the recent report published by BCC research (www.bccresearch. com), the global nutraceutical market is expected to cross $285.0 billion by 2021 from $198.7 billion in 2016 at a compound annual growth rate of 7.5%, during 5 years (Analysts, 2008). Over 1 million peoples are expected to be of age over 60 years by the year 2021 and 75% of which will be from developed countries. Therefore, the demand of anti-aging and age-defying products will be higher and most likely to hit the market of botanical drugs as herbal and side effect less solution. Apart from that the increasing cost of currently available treatment for various conditions including aging is main driving force for the patient to move toward botanical products, which are relatively safe and more effective. However, various problems such as regulatory hurdles, long approval processes and complex characterization procedures are forming barriers for the complete development of botanical products (Anis et al., 2017).

Particularly in Europe, enhanced regulatory tightness and long approval processes stifle the botanical products to become the part of commercial market. The focus of European companies is more on rebranding and expanding the available active compounds. In Latin America, undeserved markets could spur development in the region as disposable income increases. India, China, and Japan are expected to be the major consumer of botanical drugs by 2020. The reason behind the statement is the rapidly growing middle class and growing cost of current treatment of various diseases in these nations (Girdhar et al., 2017). The major leading products of natural origin are tabulated in Table 5.2.

Table 5.2 Various Natural/Botanical Products Available in Commercial Market

Class	Content	Brand Name	Company Name
Calcium supplement	Calcium and vitamins	Calcirol D3	Cadila healthcare, India
		Coral calcium	Nature's answer, USA
Protein supplement	Proteins, vitamin B	Threptin®	Raptakos, Brett and Co. India
	Proteins vitamins, minerals	Proteinex®	Pfizer, USA
Immunobooster	Amla, Pippali, Ashwagandha	Chyawanprash	Dabur India
Immune supplement	Lycopene, vitamins	Omega woman	Wassen, UK
	Dry fruit extract	Celestial healthtone	Celestial Biolabs
Nutritional supplement	Antioxidants	Selltoc AC	Fourrts, India
	Ginseng, vitamins	Revital®	Ranbaxy, India
	Proteins vitamins, minerals	GRD	Zydus cadila, India

5.6 Classification of nutraceuticals—based on natural products

5.6.1 Plants

Up until now, plants have been the largest source of drugs for curing many diseases. The magnitude of their contribution toward healthcare has been huge, and ethnobotanical references have been a guide in understanding and studying in depth about the uses of these plants. All the traditional systems of medicines have vouched for plants and even today people around the world, especially in developing countries, lean on plants and herbal remedies for the treatment of disease (Gurib-Fakim, 2006). Eleven percent of 252 drugs that are deemed vital by WHO are from plant origin. Plants also have been considered a rather difficult source to explore due to presence of complex mixtures of substances, but the rise in the hyphenated techniques have been useful to provide ease in isolating molecules from them (Kala, 2017).

Looking from the perspective of diseases, more than 3,000 plants have been used in the treatment of cancer (Chouhan et al., 2016). Some of the examples of anticancer plants are *Catharanthus roseus, Taxus brevifolia*, Podophyllum species, *Origanum majorana, Magnolia champaca*, and the molecules obtained from them are taxol, camptothecin, vincristine, vinblastine, and more (Zishan et al., 2017). If we consider antibiotics and antimicrobials, even though one of the greatest discovery is from microbial source, plants also have shown good potential as their origin e.g. *Alcea rosea, Cinnamomum cassia, Commiphora molmol, Allium sativum* and so on (Busia, 2016). As for Alzheimer's, which is a neurodegenerative disorder, plants used are ginseng species, *Crocus sativus, Ginkgo biloba, Petroselinum crispum, Nelumbo nucifera*, and they have given molecules like salvianolic A, salvianolic B, crocin, galantamine, rivastigmine, and so on (Yang et al., 2016). Many plants like *Curcuma longa, Allium sativum* and *Embelica officinalis* are known for multiple activities, which will be discussed further in this chapter.

5.6.2 Animals

People don't regard animals as a source of drugs, but they are unaware that they are an origin of many vaccines and anti-neoplastic agent. Consumption of animals is prohibited in many religions; hence medicine derived from it won't be used by the masses (de Roode et al., 2013). Sheep thyroid is a source of thyroxine used in hypertension. Cod liver oil is used as a source of Vitamin A, D, E, and K. Heparin, which is a major anticoagulant, is obtained from porcine and porcine also acts as a source of digestive supplements amylase, lipase, and protease. Animal cells are also used as vessels in cell culture (Lentjes et al., 2014). Premarin is a conjugated estrogenic product that comes from horse urine and is used in natural hormonal therapy. Insulin is obtained from the pancreas of sheep and pigs (Hubbard and Chaudry, 2013).

5.6.3 Minerals

Minerals are very important for performing various functions in the body including metabolic and non-metabolic functions. They have been entrenched for their nutritional value. They can be used to decrease the chances of anaemia, osteoporosis and to shape up strong bones and teeth. One of their important function in their involvement in forming binding site of metalloenzyme, which helps in performing essential function of the body (Kishi et al., 1996).

5.6.3.1 Iodine

Iodine deficiency disorders have been prevalent mostly in parts of Asia. It leads to stillbirths, congenital anomalies, and damaged mental functions in both adult and children. This is because deficiency of iodine has a direct impact on synthesis of thyroid hormones in the body ehich are essential to perform many bodily functions (Bouga et al., 2016). Inclusion of products containing iodine such as iodised salt and iodised oil may help in tackling the effects of hypothyroidism. Even in market there are many products containing iodine, sold in tablets and powder form that might help alleviate the effects. From nature iodine can be found from dried seaweed, cod, milk, fish and fish products, and more (Pehrsson et al., 2016).

5.6.3.2 Magnesium

It is essential to manage nerve and muscle functioning in the body and is required for more than 300 biochemical functions in the body as a co-factor for enzymes. It is required for controlling blood glucose and alleviate diabetes and to decrease the fluctuations in blood pressure. Magnesium is also required for regular cellular functions such as glycolysis, transporting calcium and potassium

Table 5.3 Examples of Various Sources of Iron

Iron Sources	Examples
Vegetables	Spinach, fruits, tomato, asparagus, apricots, beets, peas, pumpkin
Grains and dry fruits	Oats, cashew, almonds and hazelnuts
Dairy	Milk and yoghurt
Meat	All kinds of meat including duck, beef, lamb, chicken, pork etc.

across cell membrane, oxidative phosphorylation, and more. Hence along with diabetes, it is understood that it helps us in heart ailments and blood pressure issues. It can be obtained from daily food materials like fruits, leafy vegetables, almond, cashew, soy products, legumes, and milk (Mitchell, 2016).

5.6.3.3 Iron

Iron is an essential trace element which is recycled and reused in the body. It regulates various process in the body such as oxygen and electron transport along with DNA synthesis. Its deficiency can lead to anemia, neurodegenerative disorder, weakness, and disruption of many body processes. It also has adverse effects on pregnancy and immunity (González-Reyna et al., 2016). On an average, an adult need 1–3 g of iron in the body. The source of iron in nature is presented in Table 5.3.

5.6.4 Marine

Marine is a humungous slice of earth's total biodiversity but comparatively it is much less explored as a source of drug. In last two decades many novel molecules showing potent activity have been isolated from marine source (Chuyen and Eun, 2017). The organism from which bioactive molecules have been isolated are marine algae, fungi, bacteria, toxins, corals, tunicates, sea urchins, worm's vertebrates, etc. Many compounds having complex and varied chemical structures have been isolated from it including amino acids, simple peptides, nucleosides, cytokinins, alkaloids (pyridoacridine, indole, pyrrole, isoquinoline etc.) and many more. The main problems associated with this source is acquiring the material from the source as they are sometimes situated many hundred feet below sea level. Another problem is that many of these marine species are protected due to fear of its extinction so getting ample amount of sample can be thought of as a challenge (Bosch et al., 2016).

5.7 Different manufacturing methods to isolate and purify phytochemicals

5.7.1 Counter current extraction

In counter current extraction, as the name suggests the solvent and the material to be extracted moves in a counter (opposite) direction which increases the efficiency of extraction. The drug which is to be extracted must undergo size reduction which is them employed to form slurry, then slurry is set to move in one direction whereas the menstruum is passed in reverse direction. This counter movement enhances the extraction process by maintaining the concentration gradient across the exchange membrane (Belova, 2016). Thus, the menstruum passes ahead with increased concentration, where in several parameters like flow rate and solvent ratios are required to be optimized so as to get a more efficient, less time consuming and more safer technique even at elevated temperatures. Finally, from one end the extract is collected and sent for further processing wherein from other end by product is collected and used accordingly (Hartland, 2017).

The counter current method not only is employed in extraction but also in purification of obtained extract, like in study done by Jinqian Yu et al isolated five terpenoids, three known and other 2 unknown using ultra pressure extraction followed by purification using counter current method. Using petroleum ether–ethyl acetate–methanol–water in ratios of 1:0.8:1.1:0.6, Volume by volume there by helping in yielding 35.5 mg and 12.3 mg of two unknown compounds, which were further elucidated using spectroscopic techniques to identify its structure (Yu et al., 2017).

Another intriguing study where in use of hexane was replaced with ethanol due to several advantages related to it like bio renewable, good operational security and low toxicity. In this study, multiple batches for solid-liquid extraction was employed using the same principle of counter current for extraction of rice bran using ethanol, which proved to be more feasible, and also the number of theoretical plates calculated with using hexane as solvent is less as compared to ethanol where the theoretical plates are high, thus showing an enhanced efficiency in extraction. This technique of employing ethanol using counter current principle has shown promising results when compared to conventional methods for extraction (Bessa et al., 2017).

5.7.2 SCFE extraction with SCFE integrated with HPLC

Supercritical fluid extraction is a process which involves the menstruum in supercritical state, which is obtained when the triple point is crossed by applying pressure and temperature so that there is a state of counterpoise attained between liquid and vapor state (Chougle et al., 2016). Generally such techniques are been employed for industrial scale and thus, CO_2 is the most prefered gas as it have sevral benifites related to it (Subroto et al., 2017). The critical state for CO_2 could be achieved by maintaining the pressure and temperature above 74 bars and 31°C, respectively, however, such parameters could be easily modified depending on the need or requirements by adding required modifiers. Methodology involved for extraction includes, placing the sample to be extracted in column followed by preparing optimized ratio of solvent phase by mixing suitable ingredients to achieve the desired supercritical state at desired parameters (Ge et al., 2017).

In a very recent approach, investigators employed SCFE with HPLC to utilize several extraction processes. One study done by Wu et al. (2013) where in SC-CO2 extraction of C. indicum was done, and the anti-inflammatory action of C. indicum was determined. Further in the phytochemical study approximately 35 compounds where identified using GC-MS wherein the 5 compounds, luteolin-7-glucoside, chlorogenic acid, linarin, acacetin, and luteolin where re confirmed and quantitaively determined using HPLC- Pulsed Amperometric Detector (PAD). Thus, such study shows that SCFE integrated with HPLC helps in elucidating the chemical structure followed by exploration of the activity.

Another study was done by Massimo Tacchini et al. (2017) where in supercritical fluid extract (SCFE) of *Echinacea pallida* roots was done as they tend to show antiproliferative activity due to the presence of the several acetylenic compounds, and these extracts were compared with the traditional extracts obtained from the soxhlet extraction.

HPLC–UV/DAD and HPLC–ESI–MS were used as a composition analyser. The results obtained notified that SCFE proved to be a better method for extraction of polyenes and polyacetylenes from *E. pallid* roots, thus signifying SCFE to be a treasured tool for industrial application purposes in cancer prevention (Tacchini et al., 2017).

5.7.3 Subcritical water extraction

Subcritical water extraction (SWE) is considered among the most hopeful modern extraction techniques, which could be employed for the isolation of bioactive complexes from botanical origin wherein the temperature ranges between 100°C and 374°C and pressure is maintained high enough to keep the system in liquid state (Vardanega et al., 2017). Also employing water with several unique properties like high polarity, high dielectric constant, and disproportionately high boiling point for its mass however, SWE also provides several other benefits like mild operating conditions, low process cost, environmental sustainability and short process times.

There is no such officially mentioned or commercially available SWE equipment, although it is very easy to construct one for laboratory scale. It could either be a batch operation or a continuous operation. In batch, the extraction bed is fixed and solvent flow was kept from up to down, for easy removal of analytes. In case of continuous or dynamic SWE it has following units (made of stainless steel) like, three tanks, two pumps one of the two pumps are employed for carrying out (pumping) the water and extract and the other one for flushing the tubes, one extraction vessel, an oven for heating the vessel, a heat exchanger along with a pressure restrictor to maintain the appropriate pressure in the equipment and finally a sample collector (Asl and Khajenoori, 2013).

Its application in extraction from microalgal as biomass has not been much studied that extensively, regarding to such context Awaluddin et al. (2016) investigates the capability of SWE for high yield extraction of proteins and carbohydrates from the microglial biomass. For optimising the several paramters for SWE central composite design (CCD) was employed by keeping the temperature, extraction time, biomass particle size, and microalgal biomass as variables, out of which temperature is considered to be the most vital parameter to be during the SWE of microalgal biomass. Result of this extraction showed good level of carbohydrate and protein yield under the optimized condition thereby mentioning the enhanced utility of SWE for extraction of proteins and carbohydrates.

5.7.4 Extraction of phytochemicals in solvent as oil

It is well known that water serves to be the best solvent for extraction, but it also has several disadvantages related to it like, higher critical point and also requires a severe reaction conditions and lower yields of water in-soluble bio-oil. Thus, when we use non-aqueous solvents, it shows the advantage of having lower critical point, mild reaction conditions, and having high yield of water insoluble bio-oil extraction (Dhanavath et al., 2017).

Since the solvents of petroleum origin are being restricted and strictly regulated by the Food and Drug Administration (FDA) in U.S. and evaluation, authorization, and restriction of chemicals (REACH) in Europe, thus encouraging green concepts and principles for extraction, i.e., bio-based solvents. Such bio-based solvents can be extracted from bio-renewable sources like citrus peels, rapeseed, soybean, etc., ideal substitutes which should be considered, must be non-volatile organic compounds (VOCs) that have high flash points, high dissolving power, low toxicity, and less environmental impact (Schutyser et al., 2015).

One such study indicating the use of vegetable oil for extraction done by Athanasia Goula et al. (2017) developed a novel method for extraction of carotenoids from pomegranate peels using ultrasound extraction process using different oils (sunflower oil and soy oil) as solvents. In this study authors studied several parameters that could affect the extraction process like extraction temperature, solid/oil ratio, amplitude level, and extraction time, wherein 30 minutes was found to be optimum time and peel/solvent ratio of 0.10 with 58.8% as amplitude level to get high carotenoid content extract. Thus, it established a novel, greener, and more efficient method for extraction rather than using the conventional method.

5.7.5 Ultra-sonication-assisted solvent extraction

There are several shortcomings for the traditional extraction process like high cost (almost 50% of investment in a new plant is spend on the installing the process), time consuming, less efficiency, high environmental impact, late return on investment, and yield of extract as compared to solvents and reaction time spend is comparatively low (Rezaie et al., 2015). Thus, by employing an ultrasound technique, we could easily achieve shorter extraction time, reduced organic solvent consumptions, finally reduced cost and energy further such methods can be made fully automatic thereby further reducing the time required for the entire process. Various novel techniques have been developed using ultrasound like ultrasound-assisted Soxhlet extraction, combination of ultrasound, continuous ultrasound-assisted extraction and ultrasound-assisted clevenger distillation (Habibullah and Wilfred, 2016).

One such study explains the use of ultrasonication for determining the pharmaceutical from fresh water sediments by Al-Khazrajy and Boxall (2017). In this study the pharmaceuticals were extracted using an ultrasonication method from the sediment using 2% NH_4OH in methanol then followed by 2% formic acid along with methanol and further only with methanol. The analytes obtained were then detected quantitatively using Diode Array Detector (DAD) or by tandem mass spectroscopy. Overall the authors were successful in developing a method that was way more effective and reliable for determining the level of pharmaceuticals in fresh water sediments in different regions.

Another study carried out by Hoa Thi Truong et al. (2017) where in employment of alkaline treatment assisted with ultrasound-assisted extraction process was studied as an extracting tool for extracting tocols and γ-oryzanol from rice bran in organic phase while ferulic acid in aqueous phase. The analytes obtained were quantitatively detected using HPLC and UV detection. Effect of certain parameters like alkaline treatment temperature and extraction duration was evaluated wherein at 25°C the extraction for tocols and γ-oryzanol was highest and when the temperature was increased to 80°C the extraction of ferulic acid was observed to be maximum but the concentration of γ-oryzanol showed a declining trend. Thus, authors were able to develop a more efficient and cost-effective method by employing both alkaline condition and ultrasound technique.

5.7.6 Preparative column chromatography for isolation of phytochemicals

It is a powerful technique that is being employed for isolation and purification of several chemicals, pharmaceutical compounds, biological molecules, and natural products. In case of preparative column chromatography, the only difference between the normal and preparative HPLC is that in preparative HPLC sample enters the collector after passing through the detector rest all of the principle remains same (Chua et al., 2016). Preparative column is generally being employed for isolation and purification of compounds on large scale thus, the internal diameter ranges from 1 to 10 cm wherein the column particles are of size 7 μm or more and the pump attached to the system can have a pumping rate of 10 mL/min or more, also the material from the mobile phase is recoverable (Chua et al., 2016).

In a preparative column chromatography, it has various setups as a part of their instrument like solvent reservoir section which is made of stainless steel, a pump

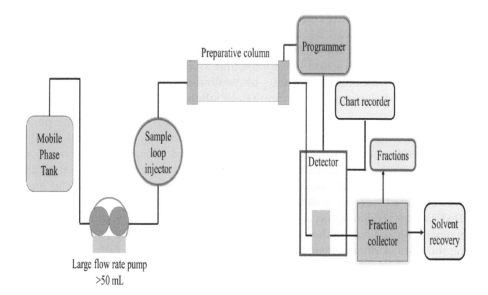

Figure 5.1 *Schematic illustration of preparative column chromatography. Solvent as mobile phase tank enters in preparative column via a sample loop injector of range 0.1 to 100 mL then passed to the detector then to fraction collector to collect the fractions when eluted or sent the solvent for recovery. Programmer can control the whole parameters of the system.*

with larger piston head to pump at a rate up to 100 mL/min, preparative injectors generally a rheodyne injector is employed, preparative columns, detectors, programmer, recycle valve, and the fraction collector that directs the flow either to the collector or to the waste using diverter valve (Li et al., 2016). The schematic illustration of preparative chromatography is shown in Figure 5.1.

The process of fraction collection again can be done as manual collection, which is based on a single plot and offers highest degree of flexibility, peak-based fraction collection which is directly based on response shown by the detector, mass-based fraction collection wherein the compound with desired mass is selectively collected, and time-based fraction collection which is based on time of interval. Therefore, depending upon need and available resources one can choose the mentioned method for fraction collection (Zhang et al., 2017).

The recent application regarding the preparative column chromatography includes a study done by Han et al. (2017) where in preparative HPLC along with counter current chromatography (CCC) was employed for separating G-quadruplex substrates from the crude extract of *Zanthoxylum ailanthoides* as a potent therapy for human cancer cells. In this study, authors used 75% ethanolic extract from the crude drug and its extract using ethyl acetate, petroleum ether, and n-butanol was done with maintained activity. For the first time three tetrahydro-protoberberines were isolated using preparative HPLC with CCC with more than 95% purity was obtained is acceptable quantity. Along with these tetrahydroprotoberberines I, II and III, Syringin, Magnoflorine and two lignin isomers (+)-lyoniresiol-3α-O-β-D-glucopyranoside and (−)-lyoniresinol-3α-O-β-d-glucopyranoside were also isolated using the same technique. Thus, compounds with higher stabilization effects on G-quadruplex were efficiently isolated with higher yield and purity.

5.7.7 Isolation and separation of phytochemicals through fermentation

Such process of employing fermentation for isolation and extraction of active constituents is done in case of various ayurvedic preparations like asava and arishta, which utilizes this technique of fermentation for generating in-situ alcohol for extraction (Koistinen et al., 2016). The majority of the process is carried out in earthen pots, but in industrial scale metal vessels, wooden vats, and porcelain jars are being employed. There are several ayurvedic formulations already available in market which is based on the same principle of fermentation based extraction examples are kanakasava, karpurasava, and dasmularista.

Also, there is being a recent trend used wherein the bio actives are fermented so as to improve its activity. An extensive review study done by Hussain et al. (2016) stated that drugs obtained from natural origin that possess several vital and essential constituents, activities of such constituents can be enhanced and the side effects related to them could be narrowed by fermenting them. As fermentation causes certain decomposition and biotransformation of the constituent's present in the herbs into a more compatible and stable form, thereby controlling the properties and altering the levels of present bioactive compounds few of the mentioned examples are listed in the Table 5.4. The mentioned study also convinces the importance and the contribution of fermentation-derived bioactive constituents utilized against various diseases like obesity and inflammation.

Another such study done by Guey-Horng Wang et al. (2016) wherein the tyrosinase inhibitory and antioxidant activity of extracts taken from Chinese herbs like Moutan Cortex Radicis (MCR), asparagus root, and walnut followed by *Bifidobacterium bifidum fermentation* at different periods. It was observed that the physiological actions of fermented extracts were considerably high when compared with non-fermented extracts, further fermented extracts were checked for cytotoxicity in CCD-966SK and B16F10 cells and found to be nontoxic (biocompatible). Thus, authors successfully showed the improved tyrosinase and antioxidant activity potentiation via fermentation of extracts thereby improving its application in as a skin-whitening agent.

Another interesting study by Juliana C. Fantinelli et al. (2017) wherein the antioxidant property of the extract obtained from Copoazu fruit was decreased and its cardioprotective property was increased. Authors generated data that clearly indicates the enhanced cardioprotective action of the Copoazu fermented extract against schemia-reperfusion damage via NOS-dependent

Table 5.4 Changes in the Contents of Bioactive Substances or Compounds after Fermentation

Herbal Medicine	Microbial Strain	Bioactive Substances/Compounds after Fermentation	References
Codonopsis lanceolata	Bifidobacterium longum, L. acidophilus, L. mesenteroides	Increased level of gallic acid, caffeic acid, vanillic acid, 4-coumaric acid, trans-ferulic acid, caffeine	Weon et al. (2013)
Rhodio larosea Lonicera japonica	Alcaligene spiechaudii	Increased level of total phenolic content	Chen et al. (2013)
Red ginseng	Phellinus linteus L. brevis M2 L. brevis	Increased Rg2, Rg3, Rg5, Rk1, CK, Rh1, Rh2, F2. Increased Rg3, Rg5, Rk1, compound K, Rh1, F2, Rg2, flavonoids. Increased Rg5, Rk1, Rg3, and F2.	Ryu et al. (2013)
Ginseng seed	B. licheniformis subtilis, Pediococcus pentosaceus, L. gasseri	Increased total sugar content, acidic polysaccharides, total phenolic compound, p-coumaric acid	Lee et al. (2015)
Radix astragali	A. oryzae M29	Increased levels of Phenolic constituents, antioxidant (3,4-di (40-hydroxyphenyl) isobutyric acid.)	Sheih et al. (2011)
Panax notoginseng	Streptococcus thermophilus L. helveticus, L. rhamnosus, L. acidophilus, B. longum, B. breve, Bifidobacterium bifidum	Increased levels of Rh1 and Rg3	Lin et al. (2010)
Oyaksungisan	Lactobacillus	Increased levels of rutin, naringin, hesperidin, poncirin.	Oh et al. (2012)
Jaeumganghwa-Tang	Lactobacillus	Increased levels of 5-HMF, paeoniflorin, glycyrrhizin, nodakenin and, nodakenetin, berberine and palmatine and hesperidin.	Kim et al. (2014)
Hwangryun-haedok-tang	Lactobacillus	Increased levels of baicalin	Shim et al. (2013)
Dioscorea zingiberensis	Trichoderma reesei	Increased level of Diosgenin	Zhu et al. (2010)
Rhizoma gastrodiae	Grifola frondosa	Increased level of Parishin and decreased levels of Gastrodigenin, p-hydroxylbenzaldehyde	Wang et al. (2013)

Source: Reproduced with permission from Hussain, A. et al., *Food Research International*, 81, 1–16, 2016.
Note: Ginsenoside metabolites Rg3, Rg5, Rk1, compound K, Rh1, F2, and Rg2 5-HMF—5 hydroxymethyl furfural.

mechanisms. Thus, such method of fermentation-based treatment of the extract to potentiate the physiological action and supressing the unwanted action is being employed widely and demands more research so that it can be employed for industrial scale.

5.8 Common botanical drug products and their therapeutic applications

5.8.1 Turmeric

Turmeric is commonly used in many households around the world. It is obtained from the dried rhizomes of *Curcuma longa* and contains important constituent's curcumin, demethoxy curcumin, and bisdemethoxycurcumin. It is useful in different activities like antiviral, antibacterial, antitumor, anti-inflammatory, antioxidant, antiulcer, anticancer, antiallergic, analgesic, and antiseptic (Kocaadam and Şanlier, 2017, Tyagi et al., 2017).

5.8.2 Ginger

Ginger is another commonly used condiment from the source Zingiber officinale. Mainly it contains volatile oil and it is made up of zingiberene and gingerol. It is used to relieve stress, as an anti-inflammatory activity, against dysmenorrhea, gastroprotective, analgesic, antibacterial, antidiabetic, rheumatoid arthritis, to tackle renal toxicity, anthelmintic, and hepatoprotective activity (Bode and Dong, 2011).

5.8.3 Liquorice

Liquorice is obtained from a legume Glycyrrhiza glabra. The word *liquorice* is derived from a Greek word which means "sweet root" Constituents present in it impart sweet taste. Triterpenoid such as glycyrrhetic acid, glycoside glycyrrhizinic acid and flavanoids like liquiritin and isoliquiritin are present in it. It is used to treat stomach ulcers, viral and microbial infections, depression, as a hepatoprotective, memory enhancer and antioxidant (Kaschubek et al., 2016).

5.8.4 Aloe

Aloe vera is a decorative plant that is found from different species of Aloe. It is grown both indoors and outdoors and has many medicinal properties. Its leaves can be differentiated into two parts—gel part and the latex part. The gel part is mostly made up of water and is applied directly on the skin to moisturise and alleviate the effects of sunburn. The latex part consists of aloin, which is the mixture of glucoside that contains barbaloin, isobarbaloin, β-barbaloin, etc. The resin present is aloesin. It also contains aloetic acid, aloesone, saponins, choline, etc. It is used for its many properties like wound healing, anti-inflammatory, antioxidant, anti-diabetic and is used to treat many bacterial infections, psoriasis, and peptic ulcer (Silva et al., 2014).

5.8.5 Fenugreek

The seeds are obtained from *Trigonella foenum-graecum* which is chiefly grown in India, Morocco, and Egypt, and is used as a spice. It contains simple pyridine-type alkaloid trigonelline, non-essential amino acid 4-hydroxyisoleucine, and an important sapogenin steroid diosgenin. It is known for activities like antihyperglycemic, antioxidant, immunomodulatory, anti-inflammatory, antipyretic, gastroprotective, and antihyperlipidemic (Gong et al., 2016).

5.8.6 Amla

Mostly all parts of the plant *Emblica officinalis* has been explored for various activities but its fruits are consumed as a part of diet by many people. This tree is considered as sacred by Hindus as it is believed to be originated from nectar of immortality. It contains constituents such as tannins, emblicanin A and B, punigluconin, and very high amounts of vitamin C. It is used as a laxative, antioxidant, antitumour, antipyretic, dyslipidemia, analgesic, antiulcer, and as a hepatoprotective (Bhandari and Kamdod, 2012).

5.8.7 Green tea

It is obtained by using leaves of *Camellia sinensis* and is consumed as a beverage all around the globe. It contains alkaloids like caffeine along with maximum amount of polyphenols such as epigallocatechin gallate, gallic acid and flavanols like quercetin and kaempferol. It contains activities like antioxidant, anticarcinogenic, ultraviolet protection, antiviral, antibacterial 2-L, and neuroprotection (Senanayake, 2013).

5.8.8 Asafoetida

The seeds of Ferula asafoetida are used as spice and are known for treating various disease such as asthma, bronchitis, and Dysmenorrhoea. It is also known for different activities such as anticancer, antimicrobial anthelminthic, antibacterial, antispasmodic, anticonvulsant, antitumour, and hypotensive agents (Vijayasteltar et al., 2017). Chemically it contains volatile oils, resins, and other constituents. Volatile oil has a pungent smell and has sulphur-containing components such as thioallyl disulphide. It contains free ferulic acid which on treatment with hydrochloric acid gives umbellic acid and umbelliferone.

5.8.9 Garlic

It is obtained from the bulbous plant *Allium sativum* and is largely cultivated in India and China. It is a part of daily diet of many people but its consumption is frowned upon in certain religions. It has a characteristic acrid and pungent taste due to the presence of volatile oil and sulphur containing compounds in it. The chemical constituents present in it are alliin, allicin, diallyl disulphide, allyl propyl disulphide. It has been known to treat colon cancer common cold, hypercholesterolemia, vaginitis, atherosclerosis, breast cancer, leukaemia, and they show immunomodulatory, antihypertensive, antimicrobial, anti-fungal, and anti-diabetic activity (Suleria et al., 2015).

5.8.10 Brahmi

Brahmi is named after Hindu god Lord Brahma, God of creation. It is obtained from *Bacopa monnieri*, which is a perennial creeping herb used in Ayurveda. It is also known as *Centella asiatica*. It contains alkaloids brahmine and herpestine along with dammarane-type triterpenoid saponins bacosides A and B and bacopaside I and II. It is used as a cognitive enhancer, neuroprotective agent, and hepatoprotective, and enhances activities such as anti-stress, anti-parkinsonism, antidiabetic, and antimicrobial.

5.8.11 Kalmegh

It is obtained from plant *Andrographis paniculata* from family Acantheceae and is used as a bitter tonic. It is cultivated in Asia and is used in Ayurvedic and Siddha traditional system of medicine. The chief constituents are cyclic diterpene lactone andrographolide and neoandrographolide. It is used as anthelmintic, hepatoprotective, immunostimulant, anticancer, and antihyperglycaemic agent (Niranjan et al., 2010).

5.8.12 Onion

It is obtained from *Allium cepa* and is used as a vegetable. It contains chemical constituents such as alliin, allicin, 2,5-dimethyl thiophene, benzyl thio-isocyanate, and is used as an anticancer, antispasmodic agent, and is used to treat respiratory problems, asthma, cardiovascular diseases, and hyperglycaemia (Cramer et al., 2014).

5.9 Regulatory status of botanical drug products

United States Food and Drug Administration issued a guidance regarding botanical drug products for industry in June 2004, which was replaced by its first revision in December 2016. These documents represent the FDA's view or thinking regarding the subject but do not impose any of it on the people (Wu et al., 2015). It includes guidelines regarding:

- General approaches
- Chemistry, manufacturing, and control
- Marketing under OTCs and NDA
- Documentation

- Clinical and non-clinical safety assessment
- Post marketing considerations

Quality control of botanical raw material and botanical drug substance and drug product.

Some other guidance documents are issued by the Union of Pure and Applied Chemistry, the EMEA and European Food Safety Authority (EFSA) (Zhu, 2017). The regulation regarding these botanical products varies from country to country, and this leads to the problems in their import and export. There is a need for international harmonization of all the laws as different information and specification of a same herb have been found in several pharmacopoeias. This affect the quality and the subsequent expected results from that herb. A common platform or organization is needed to be created to form norms and regulation controlling every aspect of these BDPs from cultivation to sale (Zhu, 2017).

5.10 Case studies related to botanical drug products

Many countries of the world are utilizing natural products as health benefit approach. Specially, the Indian, Chinese, Japanese, Korean, and Western cultures have used nature-based products since ancient times. However, well-documented investigations are lacking with several Asian botanical products (Wu et al., 2015). Complementary and alternative medicine (CAM) therapies such as traditional Chinese medicine (TCM), ayuverdic medicine, herbal therapy, acupuncture, yoga, homeopathy, chiropractic medicine, and massage therapy still gaining wide attention and are very popular as modalities for the remedy of many diseases/symptoms such as pain, asthma, cold and cough, fever, migraine, etc.

Li and Brown studied the article pertaining to the care of early age asthmatic patient using nature-based CAM techniques and concluded three clinical studies using three different combination of TCM (Li and Brown, 2009). The formula used in that studies were: altered Mai Men Dong Tang (mMMDT) (combination of five herbs), Ding Chuan Tang (DCT) (combination of nine herbs), and STA-1 (combination of ten herbs). In all three studies, the volunteers continued their prescribed medical regimen and were either given a TCM formula or a placebo. The volunteers showed enhanced forced expiratory volume (FEV1) outcome, in contrast to the volunteers administered a placebo in all three studies. All subjects were tolerant to the given TCM formulas and were safe.

Similarly, one more investigation of this type based on TCM given Sanfujiu to cure allergic condition and asthma by improving yang qi (nature of the sun; hot) in the lungs. Sanfujiu is a paste-type formulation that is composed of five Chinese medicines as previously reported by Tai et al. (2007)

This investigation used 119 participants of different age. Outcomes revealed that those suffering from asthma were more likely to report that the Sanfujiu treatment was significantly better as compared to those suffered from allergic diseases.

Stockert et al. (2007) evaluated the efficacy of a combination of laser-based therapy and probiotics as a form of treatment to cure asthma. Seventeen children were selected as subjects to take part in a randomized, placebo-controlled, double-blind investigation that treated asthmatic subjects by applying the TCM approach. Laser acupuncture was used as an alternative for needle acupuncture, and probiotics of non-pathogenic Enterococcus faecalis was given in place of the TCM herb, Jin Zhi. The untreated subjects were reported to show no improvement in asthmatic conditions while the treatment group had significantly lowered weekly peak flow variability as a measure of bronchial hyper-reactivity. The results obtained from this study was well matched with those reported by Liu and Gong who demonstrated that laser based acupuncture could also have a remarkable effect in stopping an acute asthma attack.

5.11 Future prospects and conclusion

The development of botanical drug products is moving in the right path as evident from various available drug products. However, many botanical drugs or herbal compounds have immense potential to be available commercially but issues related with their clinical, statistical, and regulatory aspects are needed to have urgent attention. As the matter of interest, the presentation of clinical data of ongoing research on botanical product development should be put forward to approval agencies with required accuracy in order to safeguard the clinical trials. From future perspectives, it is indicated that the escalating momentum to accelerate the use of botanical drug products requires well-established protocols for their approval in clinical trials. More attention should be paid to the development of appropriate statistical methodologies for determining efficacy and safety as well as toxicity of plant-based products.

Botanical products are extensively used in almost every country of the world, immaterial of traditional product of their own or an FDA-approved product. However, the understanding gap in between the "best scientific evidence" and what consumers actually use to treat a disease is required to be carefully addressed. The lack of matured and well-established protocols for identification, and characterization of chemical compounds obtained from botanical source are major causes of failure of many potential and traditional compounds to become viable, and creates hurdles in scale-up formulations of these products. Acceptance of natural remedies by the scientific community is based both on false and true premises. It is important that the study of Herbal Medicine is offered to all healthcare professionals. A collaborative effort is required from research groups and the pharmaceutical industries on producing highly effective and quality botanical products.

Acknowledgment

The authors would like to acknowledge Science and Engineering Research Board (Statutory Body Established Through an Act of Parliament: SERB Act 2008), Department of Science and Technology, Government of India for grant allocated to Dr. Tekade for research work on gene delivery and N-PDF funding (PDF/2016/003329; Dr. Maheshwari) for work on targeted cancer therapy. RT would also like to thank NIPER-Ahmedabad for providing research support for research on cancer and diabetes. The authors also acknowledge the support by Fundamental Research Grant (FRGS) scheme of Ministry of Higher Education, Malaysia to support research on gene delivery.

References

Ahmed, A., Khalid, N., Ahmad, A., Abbasi, N., Latif, M. and Randhawa, M. 2014. Phytochemicals and biofunctional properties of buckwheat: A review. *The Journal of Agricultural Science*, 152, 349–369.

Al-Khazrajy, O. S. and Boxall, A. B. 2017. Determination of pharmaceuticals in freshwater sediments using ultrasonic-assisted extraction with SPE clean-up and HPLC-DAD or LC-ESI-MS/MS detection. *Analytical Methods*, 9, 4190–4200.

Alabri, T. H. A., Al Musalami, A. H. S., Hossain, M. A., Weli, A. M. and Al-Riyami, Q. 2014. Comparative study of phytochemical screening, antioxidant and antimicrobial capacities of fresh and dry leaves crude plant extracts of Datura metel L. *Journal of King Saud University: Science*, 26, 237–243.

Allen, T. 2013. *Particle Size Measurement*. New York: Springer.

Analysts, G. I. 2008. *Nutraceuticals: Global Markets and Processing Technologies*. Denver, CO: BBC Reasearch.

Anis, M., Altaher, G., Sarhan, W. and Elsemary, M. 2017. *Nutraceutical Industry:Nanovate*. New York City: Springer.

Ansari, S. 2016. Globalization of herbal drugs. Ukaaz Publications 16-11-511/D/408, Shalivahana Nagar, Moosarambagh, Hyderabad, 500036, India.

Asl, A. H. and Khajenoori, M. 2013. Subcritical water extraction. *Mass Transfer-Advances in Sustainable Energy and Environment Oriented Numerical Modeling*. doi:10.5772/54993

Avio, C. G., Gorbi, S. and Regoli, F. 2015. Experimental development of a new protocol for extraction and characterization of microplastics in fish tissues: First observations in commercial species from Adriatic Sea. *Marine Environmental Research*, 111, 18–26.

Awaluddin, S., Thiruvenkadam, S., Izhar, S., Hiroyuki, Y., Danquah, M. K. and Harun, R. 2016. Subcritical water technology for enhanced extraction of biochemical compounds from chlorella vulgaris. *BioMed Research International*, 2016.

Azmir, J., Zaidul, I., Rahman, M., Sharif, K., Mohamed, A., Sahena, F., Jahurul, M., Ghafoor, K., Norulaini, N. and Omar, A. 2013. Techniques for extraction of bioactive compounds from plant materials: A review. *Journal of Food Engineering*, 117, 426–436.

Bakoğlu, A., Kökten, K. and Kilic, O. 2017. Seed fatty acid composition of some Fabaceae taxa from Turkey, a chemotaxonomic approach. *Progress in Nutrition*, 19, 86–91.

Belova, V. 2016. On rare metal separation by counter-current extraction in chromatography mode. *Russian Journal of Inorganic Chemistry*, 61, 1601–1608.

Bessa, L. C., Ferreira, M. C., Rodrigues, C. E., Batista, E. A. and Meirelles, A. J. 2017. Simulation and process design of continuous countercurrent ethanolic extraction of rice bran oil. *Journal of Food Engineering,* 202, 99–113.

Bhandari, P. and Kamdod, M. 2012. Emblica officinalis (Amla): A review of potential therapeutic applications. *International Journal of Green Pharmacy,* 6, 257.

Bode, A. M. and Dong, Z. 2011. The amazing and mighty ginger. *Herbal Medicine: Biomolecular and Clinical Aspects,* 2.

Bosch, A. C., O'Neill, B., Sigge, G. O., Kerwath, S. E. and Hoffman, L. C. 2016. Heavy metals in marine fish meat and consumer health: A review. *Journal of the Science of Food and Agriculture,* 96, 32–48.

Bouga, M., Lean, M. and Combet, E. 2016. MON-P248: Dietary guidance during pregnancy and iodine nutrition: A qualitative approach. *Clinical Nutrition,* 35, S244.

Brown, P. 2016. Use of chemical profiling with chemometric analyses for authentication of botanical products. *Planta Medica,* 82, OA26.

Busia, K. 2016. *Fundamentals of Herbal Medicine: Major Plant Families, Analytical Methods, Materia Medica,* Bloomington, IN: Xlibris Corporation.

Camero, C. M., Temraz, A., Braca, A. and De Leo, M. 2017. Phytochemical study of Joannesia princeps Vell.(Euphorbiaceae) leaves. *Biochemical Systematics and Ecology,* 70, 69–72.

Chandrasekara, A., Rasek, O. A., John, J. A., Chandrasekara, N. and Shahidi, F. 2016. Solvent and extraction conditions control the assayable phenolic content and antioxidant activities of seeds of black beans, canola and millet. *Journal of the American Oil Chemists' Society,* 93, 275–283.

Chen, X. C., Ren, K. F., Zhang, J. H., Li, D. D., Zhao, E., Zhao, Z. J., Xu, Z. K. and Ji, J. 2015. Humidity-triggered self-healing of microporous polyelectrolyte multilayer coatings for hydrophobic drug delivery. *Advanced Functional Materials,* 25, 7470–7477.

Chen, Y.-S., Liou, H.-C. and Chan, C.-F. 2013. Tyrosinase inhibitory effect and antioxidative activities of fermented and ethanol extracts of Rhodiola rosea and Lonicera japonica. *The Scientific World Journal,* 2013.

Chougle, J. A., Bankar, S. B., Chavan, P. V., Patravale, V. B. and Singhal, R. S. 2016. Supercritical carbon dioxide extraction of astaxanthin from Paracoccus NBRC 101723: Mathematical modelling study. *Separation Science and Technology,* 51, 2164–2173.

Chouhan, A., Karma, A., Artani, N. and Parihar, D. 2016. Overview on cancer: Role of medicinal plants in its treatment. *World Journal of Pharmacy and Pharmaceutical Sciences,* 5, 185–207.

Chua, L. S., Latiff, N. A. and Mohamad, M. 2016. Reflux extraction and cleanup process by column chromatography for high yield of andrographolide enriched extract. *Journal of Applied Research on Medicinal and Aromatic Plants,* 3, 64–70.

Chuyen, H. V. and Eun, J.-B. 2017. Marine carotenoids: Bioactivities and potential benefits to human health. *Critical Reviews in Food Science and Nutrition,* 57, 2600–2610.

Cramer, C. S., Singh, N., Kamal, N. and Pappu, H. R. 2014. Screening onion plant introduction accessions for tolerance to onion thrips and iris yellow spot. *HortScience,* 49, 1253–1261.

De Roode, J. C., Lefèvre, T. and Hunter, M. D. 2013. Self-medication in animals. *Science,* 340, 150–151.

Dhanavath, K. N., Shah, K., Bankupalli, S., Bhargava, S. K. and Parthasarathy, R. 2017. Derivation of optimum operating conditions for the slow pyrolysis of Mahua press seed cake in a fixed bed batch reactor for bio–oil production. *Journal of Environmental Chemical Engineering,* 5, 4051–4063.

Fantinelli, J. C., Álvarez, L. N. C., Arbeláez, L. F. G., Pardo, A. C., García, P. L. G., Schinella, G. R. and Mosca, S. M. 2017. Acute treatment with copoazú fermented extract ameliorates myocardial ischemia-reperfusion injury via eNOS activation. *Journal of Functional Foods,* 34, 470–477.

Ge, Y., Guo, P., Xu, X., Chen, G., Zhang, X., Shu, H., Zhang, B., Luo, Z., Chang, C. and Fu, Q. 2017. Selective analysis of aristolochic acid I in herbal medicines by dummy molecularly imprinted solid-phase extraction and HPLC. *Journal of Separation Science*, 40, 2791–2799.

Girdhar, S., Pandita, D., Girdhar, A. and Lather, V. 2017. Safety, quality and regulatory aspects of nutraceuticals. *Applied Clinical Research, Clinical Trials and Regulatory Affairs*, 4, 36–42.

Gong, J., Fang, K., Dong, H., Wang, D., Hu, M. and Lu, F. 2016. Effect of fenugreek on hyperglycaemia and hyperlipidemia in diabetes and prediabetes: A meta-analysis. *Journal of Ethnopharmacology*, 194, 260–268.

González-Reyna, A., Naranjo-García, F., Zárate-Fortuna, P., Juárez-Félix, J., Ibarra-Hinojosa, M., Limas-Martínez, A. and Martínez-González, J. 2016. Serum zinc, iron and copper in hair sheep with parenteral supplementation of minerals. *Revista de Investigaciones Veterinarias del Perú (RIVEP)*, 27, 706–714.

Goula, A. M., Ververi, M., Adamopoulou, A. and Kaderides, K. 2017. Green ultrasound-assisted extraction of carotenoids from pomegranate wastes using vegetable oils. *Ultrasonics Sonochemistry*, 34, 821–830.

Gurib-Fakim, A. 2006. Medicinal plants: Traditions of yesterday and drugs of tomorrow. *Molecular Aspects of Medicine*, 27, 1–93.

Habibullah and Wilfred, C. D. Ultrasonic-assisted extraction of essential oil from Botryophora geniculate using different extracting solvents. *AIP Conference Proceedings*, 2016. AIP Publishing, 040004.

Han, T., Cao, X., Xu, J., Pei, H., Zhang, H. and Tang, Y. 2017. Separation of the potential G-quadruplex ligands from the butanol extract of Zanthoxylum ailanthoides Sieb. and Zucc. by countercurrent chromatography and preparative high performance liquid chromatography. *Journal of Chromatography A*. 1507, 104–114.

Hartland, S. 2017. *Counter-current Extraction: An Introduction to the Design and Operation of Counter-current Extractors*, Amsterdam, the Netherlands: Elsevier.

Hubbard, W. J. and Chaudry, I. H. 2013. *Evaluation of Premarin in a Rat Model of Mild and Severe Hemorrhage*. Birmingham, UK: Alabama University.

Hussain, A., Bose, S., Wang, J.-H., Yadav, M. K., Mahajan, G. B. and Kim, H. 2016. Fermentation, a feasible strategy for enhancing bioactivity of herbal medicines. *Food Research International*, 81, 1–16.

Julier, A. C., Jardine, P. E., Coe, A. L., Gosling, W. D., Lomax, B. H. and Fraser, W. T. 2016. Chemotaxonomy as a tool for interpreting the cryptic diversity of Poaceae pollen. *Review of Palaeobotany and Palynology*, 235, 140–147.

Kala, C. 2017. Traditional health care systems and herbal medicines. *European Journal of Environment and Public Health*, 1, 03.

Kaschubek, T., Mayer, E., Schatzmayr, G. and Teichmann, K. 2016. Liquorice as a feed additive–a health promoting natural sweetener. *Planta Medica*, 81, P1003.

Kim, Y., You, Y., Yoon, H.-G., Lee, Y.-H., Kim, K., Lee, J., Kim, M. S., Kim, J.-C. and Jun, W. 2014. Hepatoprotective effects of fermented curcuma longa L. on carbon tetrachloride-induced oxidative stress in rats. *Food Chemistry*, 151, 148–153.

Kishi, K., Someren, M. K., Mayfield, M. B., Sun, J., Loehr, T. M., Gold, M. H. 1996. Characterization of manganese(II) binding site mutants of manganese peroxidase. *Biochemistry*, 35, 8986–8994.

Kocaadam, B. and Şanlier, N. 2017. Curcumin, an active component of turmeric (Curcuma longa), and its effects on health. *Critical Reviews in Food Science and Nutrition*, 57, 2889–2895.

Koistinen, V. M., Katina, K., Nordlund, E., Poutanen, K. and Hanhineva, K. 2016. Changes in the phytochemical profile of rye bran induced by enzymatic bioprocessing and sourdough fermentation. *Food Research International*, 89, 1106–1115.

Kumar Tekade, R., Gs Maheshwari, R., Sharma, P.A., Tekade, M. and Singh Chauhan, A. 2015. siRNA therapy, challenges and underlying perspectives of dendrimer as delivery vector. *Current Pharmaceutical Design*, 21, 4614–4636.

Lalu, L., Tambe, V., Pradhan, D., Nayak, K., Bagchi, S., Maheshwari, R., Kalia, K. and Tekade, R. K. 2017. Novel nanosystems for the treatment of ocular inflammation: Current paradigms and future research directions. *Journal of Controlled Release*, 268, 19–39.

Lee, H. S., Lee, Y. J., Chung, Y. H., Nam, Y., Kim, S. T., Park, E. S., Hong, S. M., Yang, Y. K., Kim, H.-C. and Jeong, J. H. 2015. Beneficial effects of red yeast rice on high-fat diet-induced obesity, hyperlipidemia, and fatty liver in mice. *Journal of Medicinal Food*, 18, 1095–1102.

Lentjes, M. A., Welch, A. A., Mulligan, A. A., Luben, R. N., Wareham, N. J. and Khaw, K.-T. 2014. Cod liver oil supplement consumption and health: Cross-sectional results from the EPIC-Norfolk Cohort Study. *Nutrients*, 6, 4320–4337.

Li, A., Xuan, H., Sun, A., Liu, R. and Cui, J. 2016. Preparative separation of polyphenols from water-soluble fraction of Chinese propolis using macroporous absorptive resin coupled with preparative high performance liquid chromatography. *Journal of Chromatography B*, 1012, 42–49.

Li, X.-M. and Brown, L. 2009. Efficacy and mechanisms of action of traditional Chinese medicines for treating asthma and allergy. *Journal of Allergy and Clinical Immunology*, 123, 297–306.

Lin, Y.-W., Mou, Y.-C., Su, C.-C. and Chiang, B.-H. 2010. Antihepatocarcinoma activity of lactic acid bacteria fermented Panax notoginseng. *Journal of Agricultural and Food Chemistry*, 58, 8528–8534.

Maheshwari, R. G., Tekade, R. K., Sharma, P. A., Darwhekar, G., Tyagi, A., Patel, R. P. and Jain, D. K. 2012. Ethosomes and ultradeformable liposomes for transdermal delivery of clotrimazole: A comparative assessment. *Saudi Pharmaceutical Journal*, 20, 161–170.

Maheshwari, R., Tekade, M., Sharma, P.A. and Kumar Tekade, R. 2015a. Nanocarriers assisted siRNA gene therapy for the management of cardiovascular disorders. *Current Pharmaceutical Design*, 21, 4427–4440.

Maheshwari, R. G., Thakur, S., Singhal, S., Patel, R. P., Tekade, M. and Tekade, R. K. 2015b. Chitosan encrusted nonionic surfactant based vesicular formulation for topical administration of ofloxacin. *Science of Advanced Materials*, 7, 1163–1176.

Mariod, A. A. 2016. Extraction, purification, and modification of natural polymers. *Natural Polymers*. New York: Springer. Pp. 63–91.

Meglič, S. H., Levičnik, E., Luengo, E., Raso, J. and Miklavčič, D. *The Effect of Temperature on Protein Extraction by Electroporation and on Bacterial Viability. 1st World Congress on Electroporation and Pulsed Electric Fields in Biology, Medicine and Food and Environmental Technologies*, 2016. New York: Springer, 175–178.

Misra, A. and Srivastava, S. 2016. Chemotaxonomy: An approach for conservation and exploration of industrially potential medicinal plants. *Journal of Pharmacognosy and Natural Products*, 2, e108.

Mitchell, H. L. 2016. Magnesium It's role in reducing sodium in foods and in the balance of minerals for heart health. *Agro Food Industry Hi-Tech*, 27, 18–22.

Niranjan, A., Tewari, S. and Lehri, A. 2010. Biological activities of kalmegh, *Andrographis paniculata Nees. and its Active Principles-A*, 1(2), 125–135.

Oh, Y.-C., Cho, W.-K., Oh, J. H., Im, G. Y., Jeong, Y. H., Yang, M. C. and MA, J. Y. 2012. Fermentation by Lactobacillus enhances anti-inflammatory effect of Oyaksungisan on LPS-stimulated RAW 264.7 mouse macrophage cells. *BMC Complementary and Alternative Medicine*, 12, 17.

Pehrsson, P. R., Patterson, K. Y., Spungen, J. H., Wirtz, M. S., Andrews, K. W., Dwyer, J. T. and Swanson, C. A. 2016. Iodine in food-and dietary supplement–composition databases. *The American Journal of Clinical Nutrition*, 104, 868S–876S.

Rezaie, M., Farhoosh, R., Iranshahi, M., Sharif, A. and Golmohamadzadeh, S. 2015. Ultrasonic-assisted extraction of antioxidative compounds from Bene (Pistacia atlantica subsp. mutica) hull using various solvents of different physicochemical properties. *Food Chemistry*, 173, 577–583.

Rosuman, P. F. and Lirio, L. G. Alkaloids as taxonomic marker of four selected species of eupatorium weeds. http://citeseerx.ist.psu.edu/viewdoc/summary?doi=10.1.1.674.4384, accessed on September 17, 2018.

Ryu, J. S., Lee, H. J., Bae, S. H., Kim, S. Y., Park, Y., Suh, H. J. and Jeong, Y. H. 2013. The bioavailability of red ginseng extract fermented by Phellinus linteus. *Journal of Ginseng Research,* 37, 108.

Sánchez-Salcedo, E. M., Mena, P., García-Viguera, C., Martínez, J. J. and Hernández, F. 2015. Phytochemical evaluation of white (Morus alba L.) and black (Morus nigra L.) mulberry fruits, a starting point for the assessment of their beneficial properties. *Journal of Functional Foods,* 12, 399–408.

Schutyser, W., Van Den Bosch, S., Renders, T., De Boe, T., Koelewijn, S.-F., Dewaele, A., Ennaert, T., Verkinderen, O., Goderis, B. and Courtin, C. 2015. Influence of bio-based solvents on the catalytic reductive fractionation of birch wood. *Green Chemistry,* 17, 5035–5045.

Senanayake, S. N. 2013. Green tea extract: Chemistry, antioxidant properties and food applications: A review. *Journal of Functional Foods,* 5, 1529–1541.

Sengar, A., Sengar, A., Sharma, V., Sharma, V., Agrawal, R. and Agrawal, R. 2017. Market development through integrating value chains: A case of Patanjali Food and Herbal Park. *Emerald Emerging Markets Case Studies,* 7, 1–22.

Sharma, P., Joshi, H., Abdin, M., Kharkwal, A. C. and Varma, A. 2017. Analytical techniques to assess medicinal plants value addition after microbial associations. *Modern Tools and Techniques to Understand Microbes.* New York: Springer.

Sharma, P.A., Maheshwari, R., Tekade, M. and Kumar Tekade, R. 2015. Nanomaterial based approaches for the diagnosis and therapy of cardiovascular diseases. *Current Pharmaceutical Design,* 21, 4465–4478.

Sheih, I.-C., Fang, T. J., Wu, T.-K., Chang, C.-H. and Chen, R.-Y. 2011. Purification and properties of a novel phenolic antioxidant from radix astragali fermented by Aspergillus oryzae M29. *Journal of Agricultural and Food Chemistry,* 59, 6520–6525.

Shim, K.-S., Kim, T., Ha, H., Lee, K. J., Cho, C.-W., Kim, H. S., Seo, D.-H. and Ma, J. Y. 2013. Lactobacillus fermentation enhances the inhibitory effect of Hwangryun-haedok-tang in an ovariectomy-induced bone loss. *BMC Complementary and Alternative Medicine,* 13, 106.

Silva, R. M. D., Ribeiro, R. D. T. M., Souza, R. J. C. D., Oliveira, A. F. M. D., Silva, S. I. D. and Gallão, M. I. 2017. Cuticular n-alkane in leaves of seven neotropical species of the family Lecythidaceae: A contribution to chemotaxonomy. *Acta Botanica Brasilica,* 31(1).

Silva, S., Oliveira, M. B., Mano, J. and Reis, R. 2014. Bio-inspired Aloe vera sponges for biomedical applications. *Carbohydrate Polymers,* 112, 264–270.

Soni, N., Soni, N., Pandey, H., Maheshwari, R., Kesharwani, P. and Tekade, R. K. 2016. Augmented delivery of gemcitabine in lung cancer cells exploring mannose anchored solid lipid nanoparticles. *Journal of Colloid and Interface Science,* 481, 107–116.

Soni, N., Tekade, M., Kesharwani, P., Bhattacharya, P., Maheshwari, R., Dua, K., Hansbro, P. and Tekade, R. 2017. Recent advances in oncological submissions of dendrimer. *Current Pharmaceutical Design,* 23(21), 3084–3098.

Srivastava, S., Misra, A., Mishra, P., Shukla, P., Kumar, M., Sundaresan, V., Negi, K. S., Agrawal, P. K. and Rawat, A. K. S. 2017. Molecular and chemotypic variability of forskolin in coleus forskohlii Briq., a high value industrial crop collected from western himalayas (India). *RSC Advances,* 7, 8843–8851.

Stockert, K., Schneider, B., Porenta, G., Rath, R., Nissel, H. and Eichler, I. 2007. Laser acupuncture and probiotics in school age children with asthma: A randomized, placebo-controlled pilot study of therapy guided by principles of Traditional Chinese Medicine. *Pediatric Allergy and Immunology,* 18, 160–166.

Subroto, E., Widjojokusumo, E., Veriansyah, B. and Tjandrawinata, R. R. 2017. Supercritical CO_2 extraction of candlenut oil: Process optimization using Taguchi orthogonal array and physicochemical properties of the oil. *Journal of Food Science and Technology,* 54, 1286–1292.

Suleria, H. A. R., Butt, M. S., Khalid, N., Sultan, S., Raza, A., Aleem, M. and Abbas, M. 2015. Garlic (Allium sativum): Diet based therapy of 21st century—A review. *Asian Pacific Journal of Tropical Disease*, 5, 271–278.

Tacchini, M., Spagnoletti, A., Brighenti, V., Prencipe, F. P., Benvenuti, S., Sacchetti, G. and Pellati, F. 2017. A new method based on supercritical fluid extraction for poly-acetylenes and polyenes from echinacea pallida (Nutt.) nutt. roots. *Journal of Pharmaceutical and Biomedical Analysis,* 146, 1–6.

Tai, C.-J., Chang, C.-P., Huang, C.-Y. and Chien, L.-Y. 2007. Efficacy of Sanfujiu to treat aller-gies: Patient outcomes at 1 year after treatment. *Evidence-Based Complementary and Alternative Medicine*, 4, 241–246.

Tekade, R. K., Maheshwari, R., Soni, N. and Tekade, M. 2017a. Chapter 12: Carbon nano-tubes in targeting and delivery of drugs A2—Mishra, Vijay. In: Kesharwani, P., Amin, M. C. I. M. and Iyer, A. (eds.) *Nanotechnology-Based Approaches for Targeting and Delivery of Drugs and Genes.* Cambridge MA: Academic Press.

Tekade, R. K., Maheshwari, R., Soni, N., Tekade, M. and Chougule, M. B. 2017b. Chapter 1: Nanotechnology for the Development of Nanomedicine A2—Mishra, Vijay. In: Kesharwani, P., Amin, M. C. I. M. and Iyer, A. (eds.) *Nanotechnology-Based Approaches for Targeting and Delivery of Drugs and Genes.* Cambridge MA: Academic Press.

Tekade, R. K., Maheshwari, R. and Tekade, M. 2017c. 4: Biopolymer-based nanocompos-ites for transdermal drug delivery. *Biopolymer-Based Composites.* Sawston, CA: Woodhead Publishing.

Tekade, R. K., Maheshwari, R., Tekade, M. and Chougule, M. B. 2017d. Chapter 8: Solid lipid nanoparticles for targeting and delivery of drugs and genes A2—Mishra, Vijay. In: Kesharwani, P., Amin, M. C. I. M. and Iyer, A. (eds.) *Nanotechnology-Based Approaches for Targeting and Delivery of Drugs and Genes.* Cambridge, MA: Academic Press.

Tomita, K., Machmudah, S., Fukuzato, R., Kanda, H., Quitain, A. T., Sasaki, M. and Goto, M. 2014. Extraction of rice bran oil by supercritical carbon dioxide and solubility consideration. *Separation and Purification Technology,* 125, 319–325.

Truong, H. T., Luu, P. D., Imamura, K., Takenaka, N., Matsubara, T., Takahashi, H., Luu, B. V. and Maeda, Y. 2017. Binary solvent extraction of tocols, γ-oryzanol, and ferulic acid from rice bran using alkaline treatment combined with ultrasonication. *Journal of Agricultural and Food Chemistry.* 65, 4897–4904.

Tyagi, A. K., Prasad, S., Majeed, M. and Aggarwal, B. B. 2017. Calebin A, a novel compo-nent of turmeric, suppresses NF-κB regulated cell survival and inflammatory gene products leading to inhibition of cell growth and chemosensitization. *Phytomedicine,* 34, 171–181.

Van Wyk, B.-E. and Wink, M. 2017. Medicinal plants of the world. *Medicinal Plants of the World.* Oxfordshire, UK: CABI.

Vardanega, R., Carvalho, P. I., Santos, D. T. and Meireles, M. A. A. 2017. Obtaining prebiotic carbohydrates and beta-ecdysone from Brazilian ginseng by subcritical water extrac-tion. *Innovative Food Science and Emerging Technologies,* 42, 73–82.

Vijayasteltar, L., Jismy, I., Joseph, A., Maliakel, B., Kuttan, R. and Krishnakumar, I. 2017. Beyond the flavor: A green formulation of Ferula asafoetida oleo-gum-resin with fenugreek dietary fibre and its gut health potential. *Toxicology Reports,* 4, 382–390.

Villarreal, M. 2017. *Value Added Products Utilizing Acid Whey: Development of a Fruit Yogurt Beverage and a Sports Drink.* Ithaca, NY: Cornell University.

Wang, G.-H., Chen, C.-Y., Lin, C.-P., Huang, C.-L., Lin, C.-H., Cheng, C.-Y. and Chung, Y.-C. 2016. Tyrosinase inhibitory and antioxidant activities of three Bifidobacterium bifidum-fermented herb extracts. *Industrial Crops and Products,* 89, 376–382.

Wang, N., Wu, T.-X., Zhang, Y., Xu, X.-B., Tan, S. and Fu, H.-W. 2013. Experimental analysis on the main contents of Rhizoma gastrodiae extract and inter-transformation through-out the fermentation process of Grifola frondosa. *Archives of Pharmacal Research,* 36, 314–321.

Weon, J. B., Yun, B.-R., Lee, J., Eom, M. R., Ko, H.-J., Kim, J. S., Lee, H. Y., Park, D.-S., Chung, H.-C. and Chung, J. Y. 2013. Effect of codonopsis lanceolata with steamed and fermented process on scopolamine-induced memory impairment in mice. *Biomolecules and Therapeutics*, 21, 405.

Wilson, A. M., Bailey, P. J., Tasker, P. A., Turkington, J. R., Grant, R. A. and Love, J. B. 2014. Solvent extraction: The coordination chemistry behind extractive metallurgy. *Chemical Society Reviews*, 43, 123–134.

Wu, K., Wu, C., Dou, J., Ghantous, H., Lee, S. and Yu, L. 2015. A update of botanical drug development in the United States: Status of applications. *Planta Medica*, 81, PG4.

Wu, X.-L., Li, C.-W., Chen, H.-M., Su, Z.-Q., Zhao, X.-N., Chen, J.-N., Lai, X.-P., Zhang, X.-J. and Su, Z.-R. 2013. Anti-inflammatory effect of supercritical-carbon dioxide fluid extract from flowers and buds of Chrysanthemum indicum Linnen. *Evidence-Based Complementary and Alternative Medicine*, 2013.

Yadav, R., Khare, R. and Singhal, A. 2017. Qualitative phytochemical screening of some selected medicinal plants of shivpuri district (MP). *International Journal of Life Sciences Research*, 3, 844–847.

Yang, L., Yang, C., Li, C., Zhao, Q., Liu, L., Fang, X. and Chen, X.-Y. 2016. Recent advances in biosynthesis of bioactive compounds in traditional Chinese medicinal plants. *Science Bulletin*, 61, 3–17.

Yu, J., Zhao, H., Wang, D., Song, X., Zhao, L. and Wang, X. 2017. Extraction and purification of five terpenoids from Olibanum by ultrahigh pressure technique and high-speed counter-current chromatography. *Journal of Separation Science*, 40, 2732–2740.

Zhang, L., Sun, A., Li, A., Kang, J., Wang, Y. and Liu, R. 2017. Isolation and purification of osthole and imperatorin from Fructus Cnidii by semi-preparative supercritical fluid chromatography. *Journal of Liquid Chromatography and Related Technologies*, 40(1–8), 407–414.

Zhu, C., Wakeham, S. G., Elling, F. J., Basse, A., Mollenhauer, G., Versteegh, G. J., Könneke, M. and Hinrichs, K. U. 2016. Stratification of archaeal membrane lipids in the ocean and implications for adaptation and chemotaxonomy of planktonic archaea. *Environmental Microbiology*, 18, 4324–4336.

Zhu, Y. 2017. European Union regulatory and quality requirements for botanical drugs and their implications for Chinese herbal medicinal products development. *Zhongguo Zhong yao za zhi= Zhongguo Zhongyao Zazhi= China Journal of Chinese Materia Medica*, 42, 2187.

Zhu, Y.-L., Huang, W., Ni, J.-R., Liu, W. and Li, H. 2010. Production of diosgenin from Dioscorea zingiberensis tubers through enzymatic saccharification and microbial transformation. *Applied Microbiology and Biotechnology*, 85, 1409–1416.

Zishan, M., Saidurrahman, S., Anayatullah, A., Azeemuddin, A., Ahmad, Z. and Hussain, M. W. 2017. Natural products used as anti-cancer agents. *Journal of Drug Delivery and Therapeutics*, 7, 11–18.

The Role of Computational Approach in R&D of Botanical Drug Products

Aniko Nagy, Markie Esmailian, and Timea Polgar

Contents

6.1 Current landscape of plant-based medicine research through the eyes of a computational scientist

6.1.1 Introduction

"Natural products," generally, is a broad term for plants, animals, minerals, microorganisms, and their metabolites [1]. The term "botanical drug" for natural products was coined in United States, referring to drugs prepared with plant substances, such as algae and microfungi (the term botanical drugs or natural products will be used as equal terms hereafter) [2]. In the European Union (EU), the term "herbal medicinal product" (HMP) is applied to all drugs that contain at least one kind of herbal substances [3]. Traditional Chinese Medicine (TCM) was developed based on the Chinese own experience [4], while Ayurveda is used in India [5].

While several product makers promote the use of botanical products as dietary supplements, only a few products have been approved by the U.S. Food and Drug Administration (FDA) as prescription drugs (Figure 6.1). Just to mention a couple of examples: the first botanical drug derived from green tea (Camellia sinensis Kuntze) approved by the FDA was Veregen®, offering a treatment for genital and perianal warts that [6], the second one was Mytesi® (Fulyzaq®), an indicator drug for HIV-associated diarrhoea. Fulyzaq is extracted from the blood-red latex of the South American croton tree (*Croton lechlerii* Müll. Arg) [7,8].

In recent years, there has been a growing interest in the application of herbal medicine for both prevention and treatment of various diseases all over the world. In China, for example, almost half of the commercial drugs are botanical drugs, which are widely applied for the treatment of various chronic diseases like cardiovascular disease and cancer.

Herbal medicine is usually composed of many herbs in certain proportion [9,10]. The constituent herbs and their proportion are determined based on traditional information. Herbal medicine, unlike modern pharmaceutical drugs as single chemical agents, mostly contains hundreds of chemical compounds. Many scientists assume and believe that synergies of the various active ingredients add to the final physiological effect [11]. Some clinical studies proved that a botanical drug composed of multiple herbs in given proportion has greater efficacy than a single herb [12,13]. Thus, modern botanical

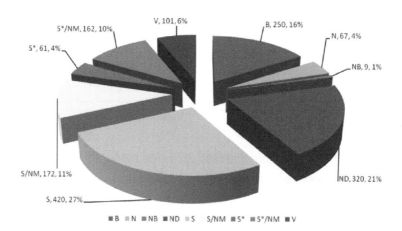

Figure 6.1 *All new approved drugs 1981–2014; (n = 1562.)(From Newman, D.J. and Cragg, G.M., J. Nat. Prod., 79, 629–661, 2014.)*

drugs can be manufactured as a combination of different active components from herbs [14].

"B" - Biological drug, usually a long peptide or protein; isolated from an organism/cell line or produced by biotechnological means in a surrogate host

"N" - Natural product, unmodified in structure, though might be semi- or totally synthetic.

"NB" - Natural product "botanical drug" (in general these have been recently approved).

"ND" - Derived from a natural product and is usually a semisynthetic modification.

"S" - Totally synthetic drug, often found by random screening/modification of an existing agent.

"S*" - Made by total synthesis, but the pharmacophore is/was from a natural product.

"V" - Vaccine.

http://pubs.acs.org/doi/abs/10.1021/acs.jnatprod.5b01055 Further permissions related to the material excerpted should be directed to the ACS.

A prominent example demonstrating chemical complexity is Cannabis, an ancient plant with new purpose. Cannabis has over 500 chemicals, some with physiological effect [15]. Moreover, scientists and medical practitioners are finding that whole-plant cannabis extracts are superior to isolates in their medical application that supports the importance of synergistic effects of active chemical components. These assumptions further support the hypothesis of the so-called "entourage effect." Tetrahydrocannabinol (THC) and cannabidiol (CBD) seem imperative to be introduced with other chemical compounds found in Cannabis.

Since 1964 THC has been the primary focus of cannabis research, when Raphael Mechoulam discovered and synthesized it. More recently, the synergistic effects of CBD to human pharmacology have been scientifically demonstrated. Other phytocannabinoids, including tetrahydrocannabivarin, cannabichromene and cannabigerol (Figure 6.2) exert other effects that can be of therapeutic interest. The terpenoids share a common substructure with phytocannabinoids, and frequently used in human diets as flavour and fragrance additives. Terpenoids are designated as "Generally Recognized as Safe" by regulatory agencies. However, terpenoids have physiological effects that can add to the entourage effect. Cannabis chemicals are under investigation for the treatment of pain, inflammation, depression, anxiety, addiction, epilepsy, cancer, fungal, and bacterial infections. Synergy between phytocannabinoid and terpenoid components can increase the likelihood an extensive pipeline of new therapeutic products from this venerable plant [16,17].

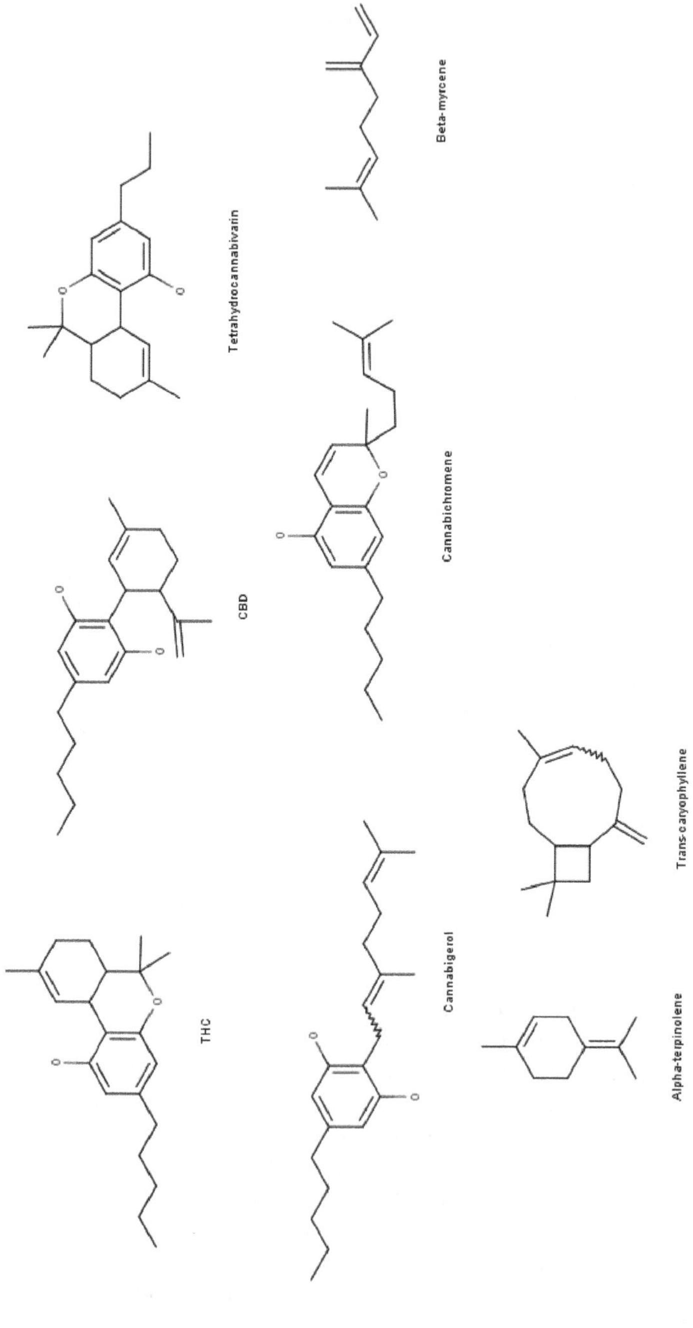

Figure 6.2 Chemical components in Cannabis.

6.1.2 Challenges
6.1.2.1 Integrated and reliable databases supporting evidence-based botanical drug research

Scientific literature analysis is the first step toward providing the justification for further studies. Moreover, botanicals that have been in use historically are often not supported by some scientific studies. Most scientific studies are *in vitro* and animal studies, case reports, or small scale, poorly designed trials further complicating the scientific data analytics. Another challenge is that many are published in foreign language. Current practices support the publication of only positive results, while the peer review process can lead to publication of papers with changing reliability. Therefore, the process of such comprehensive analysis is time consuming, often cannot be done, and calls for quality structured and standardized databases (Figure 6.3) [18–22].

Another source of information is standard pharmacopoeias that include monographs on selected botanicals. For example, the English version of the Chinese pharmacopeia has information on hundreds of Chinese medicinal

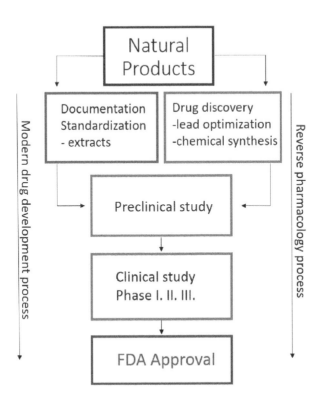

Figure 6.3 *Evidence-based botanical drug research supported by computational approaches.*

herbs [23]. The German Commission E monographs contain information on the safety and efficacy of hundreds on botanicals [24]. Pharmacopoeias are not available in structured data format therefore their analysis is also time consuming.

Ethnobotanical studies drawing attention to useful botanicals can also initiate drug discovery research [25]. The research of how various cultures use local plants can lead to valuable clues about their medicinal properties. Practitioners, however, are usually shamans or native healers who apply these herbs based on faith and spiritual beliefs. Knowledge is usually transferred through apprenticeship. To include ethnobotany, scientists must work with ethnobotanists who have a deep understanding of native culture and language. Scientists need to respect the rights of indigenous people who can claim the ownership, in addition to moral obligation to benefit of the local people. Reliable data sources are a prerequisite to ensure not only the knowledge transfer and preservation but protection of local interests. Nagoya protocol is one initiative that aims at its resolution.

Cannabis has also a long ethnobotanical history that can provide valuable insights into current research. Cannabis for medicinal use has recently become legal in the United States. Anecdotal data from medical cannabis users show promising health outcomes, but there is only limited evidence-based scientific data to validate clinical these efficacies. Challenges with U.S. federal regulations and little scientific knowledge base in the industry had made it difficult to conduct clinical trials and collect clean structured data. Lack of structured data and standardization in product development, make it challenging to generate accurate data analysis for efficacy [15].

6.1.2.2 Integrated databases supporting plant extract banks

The broadly accepted approach to drug discovery involves screening large numbers of botanicals or other agents using *in vitro* and animal models. Today, computers that can be used to simulate molecular mechanism by means of *in silico* studies are assisting these processes [26]. Botanicals that demonstrate the intended bioactivities are fractionated, then active compounds are isolated to elucidate mechanism of action. Single chemical components can be modified chemically to change or enhance the pharmacologic properties. Promising agents then are studied for toxicity, bioavailability, and gene expression, following which well characterized and purified compounds will be tested in clinical trials (Figure 6.4).

There are several botanical extract libraries that can facilitate botanical drug research. One is the Drug Development Program, DTP, of the National Cancer Institute (NCI)/National Institutes of Health (NIH), that maintains a Natural Products Repository with over 50,000 plant samples from around the world through contracts with major botanical gardens and arboretums [27]. Extensive registration systems are needed to capture the knowledge of such

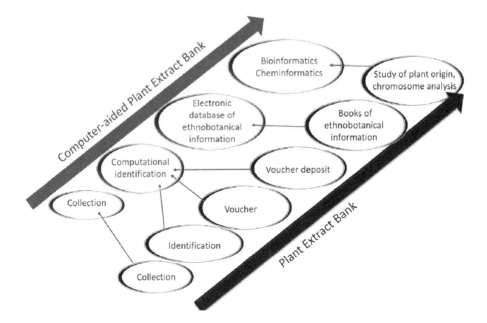

Figure 6.4 *Workflow of creating plant extract banks.*

repositories and screening campaign based on analogy to informatics systems designed for high-throughput screen analytics and chemical inventory, registration software systems.

6.1.2.3 Complex chemical composition influencing multiple targets with synergistic effects

Plant-derivatives are much complex, then chemicals in conventional single-target and single-drug development. The recent decade has proved that the single-drug "magic bullet" strategy is not adequate for applying therapy for chronic illnesses like cancers, immune disorders, mental illnesses, cardiovascular diseases, lifestyle diseases, due to their complex pharmacology. These diseases require a "synergistic multi-target and multi-component" approach involving control over a number of target sites [28].

The chemical composition of a drug determines the cellular mechanism of action and its synergistic effects, meaning that effects of some components add up while others antagonize or modulate each other's effects resulting in the final cellular output. This is a systems biological problem. Based on the chemical description, the mechanism of action could be modeled, and the resulting physiological effect could be better understood or even predicted. The traditional physiologic effects and the evidence-based (Western medical) clinical studies on several plants and products can serve as reference points for such studies [29]. The complexity of the herbal medicines calls for informatics and computational approaches that can integrate and analyse the vast amount of information and help understand the complex mechanism of action.

Reverse-engineering of traditional herbal medicine is called reverse pharmacology (RP), a current approach to unlock the pharmacology of the herbal medicine (Figure 6.5) [30,25].

RP involves the study of active ingredients based on traditional medical knowledge and formulations as well as the subsequent development of drug candidates or formulations for preclinical and clinical research. RP is essentially a transdisciplinary approach for clinical significance and for drug repurposing. The clinically novel biodynamic actions of traditional drugs may open up new doors in biomedicine and life sciences. The phyto-active molecules can provide novel chemical scaffolds for the structural modifications with drug targets. Moreover, RP can inspire new drugs from traditional remedies and enrich the chemical repertoire of medicinal chemists for novel chemical entities (NCEs). RP workflow requires the information management of the complex process: from literature surveys, understanding the molecular mechanism of the chemical mixture by computational models until data registration or data analytics of screening data or structured databases.

6.1.2.4 Proving clinical efficacy of botanicals

Modern biomedical research is hypothesis-driven, specific research questions, such as whether a single drug molecule can be used to treat a selected symptom or a disease. Traditional medicine, however, uses multiple herbs and complex treatment modalities. Therefore, the wisdom of applying modern research methodology for traditional medicine and botanicals research is often questioned. Some propose to study traditional medicine as a whole, in a so-called black box design [16]. Herbalists often personalize the formulation to each individual's unique condition. Such an approach, however, results in multiple variables, too many to control in a scientific clinical study. Moreover, its outcomes are too broad to draw meaningful conclusions and consequently has limited scientific value [31].

Today, three-phase clinical studies using a single herbal component or a well-defined formula remain the norm. First, history of use generally is employed to justify the safety of botanicals, but this does not indicate the safety of the product when used in combination with other pharmaceutical treatment. Evidence of safety should include endpoints related to interactions with other drugs that could be used in combination. As for dosage, traditional dosage range typically is applied as the therapeutic range, dose escalation trials are rarely needed. For botanicals, the dose curve may peak in a nonlinear fashion, indicating that higher dose may be less effective than a lower dose [32]. Thus it is required to determine the safety and optimal dose in a phase 1 study [33]. While in phase 2 studies, botanical agents are administered to

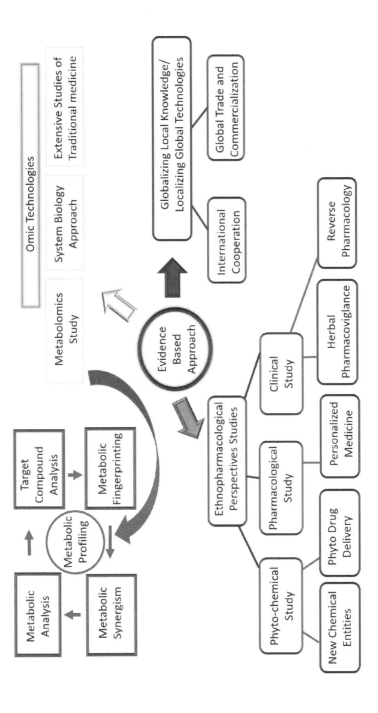

Figure 6.5 *The development of a Botanical Drug by RP needs 5–6 years, while modern drug development takes 10–15 years at significantly higher costs.*

a larger group of people to check effectiveness and to further evaluate safety. Phase 3 trials are conducted in large groups of people to confirm the effectiveness of the botanical, monitor its adverse events, and compare it to commonly used treatments. Placebos are considered robust controls in botanicals trials for several indications, when a placebo is employed, inert colouring and flavouring agents are used to give it a final appearance similar to the study product to ensure blinding.

6.1.2.5 Regulatory and administrative challenges

Research and development of Cannabis has been facing with most challenges since the discovery of THC as this substance is classified as a Schedule I of Controlled Substance in most developed countries. Despite the expansion of U.S. state-level legalizations and growing population of new users, the U.S. federal government imposes strict rules preventing research necessary to show both the harm and the benefits of the cannabis use. This poses a huge public health and safety issue [34,35].

The users in states that have stricter regulations for medical cannabis use are facing even more health and risk challenges as they consume cannabis not knowing its source, quality, and potency. National Institute on Drug Abuse (NIDA) is the only source of cannabis for research; this does not allow for varieties of cannabis people consume to be studied. The data from NIDA provided products for research has not produced clean and consistent data for analysis due to challenges associated with limited plant genetics [36].

6.2 Informatics approaches

In modern pharmaceutical research, computer-aided drug design (CADD) methods have accelerated the pace of drug discovery over the past decades [37]. Exploring the huge expanse of the theoretical chemical space has been a long-standing ambition of computer-aided drug design. The first De Novo method was shown nearly 30 years ago, that could automatically design molecules. Despite their conceptual appeal, De Novo methods have suffered from drawbacks—one is poor synthetic feasibility of the designed molecules, and another is lack of predictive power of the scoring functions. In practice, the designed molecules are generally not the ones that are synthetized; instead, they inspire medicinal chemists to carry out their own designs—in other words, the expert system is mainly the human user. Other computational methods, such as docking, also depends on the human user: Although impressive enrichment factors are often reported, they owe a large part of their success to expert intervention [38]. Structure-based drug design can deliver a 1–20% hit rate, which is ten to several hundred-fold improvement over a random high-throughput screening. Similar effectiveness rates have been reported for ligand-based approaches. Clearly, the combination of computational tools and expert users offers a very powerful solution, but it must be acknowledged that the success and ubiquity of virtual screening is due, in large part, to the continuous growth of commercial catalogues, which are easily accessible through public databases. Obtaining one true active out of every 20 molecules tested is perfectly acceptable when the compounds are relatively inexpensive and readily available. The cost is much higher when truly virtual compounds are explored, because valuable synthetic resources must be invested upfront.

In addition to ligand and structure-based CADD methods, machine-learning methods have been extensively applied to the process of drug discovery. Multiple linear regression (MLR) is among the most widely used methods to derive linear mapping between the activity and the values of structural features [39], in addition various neural networks, such as backpropagation neural networks, Bayesian regularized neural networks, and probabilistic neural networks, were also applied as valuable tools studying the structure-activity relationship of drugs.

Because detailed structural information of all compounds in herbal medicine is not always available, these methods sometime cannot be directly used to predict bioactivity of herbal medicine [40]. However, variation of biologic activity of herbal medicine is tightly associated with the variation of their chemical composition. Such relationship between chemical composition and biologic activity was regarded as quantitative composition–activity relationship (QCAR). By quantitatively analyzing the chemical composition–bioactivity relationship, the mathematical model could be established to predict activity of herbal medicine in the same manner that parallels current QSAR study. Moreover, optimal combination of herbal medicine can be calculated based on QCAR model, which enables us to integrate different active components to form a more effective botanical drug [41].

Important to note, that all computational methods rely on preexisting experimental data. Structure-based methods benefit greatly from the continuous increase in the number, diversity, and complexity of structural data. Structural genomics initiatives guarantee a continuous increase in the coverage of disease-associated proteins, while methodological advances can provide a more realistic and dynamic view of conformational heterogeneity, even in automated high-throughput applications such as fragment screening. On the ligand side, several initiatives are making it ever easier to access and use well-curated bioactivity data. Therefore, the data sources are preliminary importance of the development of novel computational methods that can be generally applied in botanical drug discovery and modeling.

6.2.1 Databases in botanical drug research

Today the Internet is the first choice of information for scientists, physicians, and other healthcare professionals. With respect to the field of botanical drug research, several investigations of commercial websites have found heterogeneous quality regarding the content presented, which might lead the scientists in a wrong direction. As a consequence, several independent initiatives have been started to assist researchers. Over the last decade, one application of modern information technology is the establishment of specialized databases. Unfortunately, information about these databases is controversial and often not known to researchers in the field. Thus, there is a basic need to give an overview on published database resources for natural product research.

Here we provide an overview and list the most relevant databases in the field (Table 6.1). In addition, we selected two representatives that we described in detail.

Most databases were created to collect the molecular content of TCM plants primarily to connect the ancient knowledge to western medicine. One of the biggest Chinese herbal database is Traditional Chinese Medicine Integrative Database (TCMID) [42], which integrate data of Chinese databases, like TCM, TCM-Taiwan and HIT (Herb Ingredients' Target) and build the relations between the herbal ingredients and their effects. Chinese databases contain the detailed information for preparation [43,44]. In addition, chemical structures of the herbs are added, which is related to diseases through targets from DrugBank [45] and OMIM databases [46]. Moreover, there are integrated data from text mining books (for example Encyclopedia of Traditional Chinese Medicines) and published articles, which are specifically valuable for Western medicine. TCMID stores records about 8159 herbs, 46914 TCM mixture, more than 25210 herb ingredients and 17500 targets.

There are six ways to query TCMID as of today:

- Prescription Query: The information of prescriptions is available in Chinese, so user can search prescriptions by using their names in Chinese pinyin.
- Herb Query: The herb can be queried with Chinese pinyin name or English name, but accepted Latin scientific name is not available for search, which would be needed to ensure the congruence and increase the adaptability of the database.
- Ingredient Query: Compound name in English, STITCH database ID, CAS number are all accepted for the query, however, the English names are not well standardized, which is disturbing for the search and modify the results.
- Disease Query: Disease information were collected from OMIM, so user can conduct the query according to OMIM phenotype name or phenotype ID.
- Drug Query: The drug English name and its Drugbank ID can be used.
- Target Query: Any kind of protein name or Uniprot AC ID is accepted (TCMID)

Table 6.1 Most Relevant Databases as References for Natural Product Research

Name	Type	Link
CHMIS-C	Herbal therapies specifically for cancer	http://sw16.im.med.umich.edu/chmis-c/
CUSTOMARY MEDICINAL KNOWLEDGE BASE	Herbal therapies specifically of Aboriginal origin	http://biolinfo.org/cmkb/index.php
EXTRACT DATABASE	Herbal therapies	http://www.plant-medicine.com/grades/extract/main-menu.asp
HERBMED & HERBMEDPRO	Herbal therapies	http://www.herbmed.org
IBIDS	Dietary supplements, vitamins, minerals, herbs, botanical and agricultural science, Phytomedicine	http://grande.nal.usda.gov/ibids/index.php
MEDFLOR INDIA/ABIM (ANNOTATED BIBLIOGRAPHY OF INDIAN MEDICINE)	Herbal therapies	http://indianmedicine.eldoc.ub.rug.nl/root/R/95965/?pFullItemRecord=0N
NAPRALERT	Herbal therapies	http://www.napralert.org/default.aspx
NATIVE AMERICAN ETHOBOTANY	Herbal therapies	http://herb.umd.umich.edu
NATURAL MEDICINES COMPREHENSIVE DATABASE	Herbal therapies, TCM	http://www.naturaldatabase.com
NATURAL STANDARD	Foods, herbs, supplements, health, wellness, brand names	http://www.naturalstandard.com
ONCORX	Traditional Chinese Medicine for cancer	http://www.onco-informatics.com/
PLANT DATABASE OF INDIA	Herbal therapies	http://221.135.191.194/plantsindia/index.php
PLANTMARKERS	Herbal therapies	http://markers.btk.fi/
PLANT	Herbal therapies	http://www.brazilian-plants.com
TCMGENEDIT	Traditional Chinese Medicine	http://tcm.lifescience.ntu.edu.tw/
TCM-INFORMATION	Traditional Chinese Medicine	http://tcm.cz3.nus.edu.sg/group/tcm-id/tcmid.asp
TCMLARS	Traditional Chinese Medicine	http://www.cintcm.com/index.htm

(Continued)

Table 6.1 (Continued) Most Relevant Databases as References for Natural Product Research

Name	Type	Link
TRADIMED	Traditional Chinese Medicine	http://www.tradimed.com
DICTIONARY OF NATURAL PRODUCTS (DNP)	Comprehensive database with synonyms, physicochemical properties	http://dnp.chemnetbase.com/faces/chemical/ChemicalSearch.xhtml
REAXYS	Comprehensive database with synonyms, physicochemical properties	https://www.reaxys.com/reaxys/session.do
SUPER NATURAL II	Largest freely available database of NPs	http://bioinf-applied.charite.de/supernatural_new/index.php
UNPD	Universal NP database	http://pkuxxj.pku.edu.cn/UNPD
TCM DATABASE@TAIWAN	Traditional Chinese Medicine	http://tcm.cmu.edu.tw
TCMID	Traditional Chinese Medicine	www.megabionet.org/tcmid
CHEM-TCM	Traditional Chinese Medicine (chemical database)	www.chemtcm.com
HIT	Traditional Chinese Medicine	http://lifecenter.sgst.cn/hit
HIM	Traditional Chinese Medicine	http://binfo.shmtu.edu.cn:8080/him
AFRODB	African natural products	http://african-compounds.org/about/afrodb/
AFROCANCER	African natural products focus on anticancer activity	http://african-compounds.org/about/afrocancer/
AFROMALARIADB	African natural products focus on antimalarial activity	http://african-compounds.org/about/afromalariadb/
SANCDB	South African Natural Compound Database	http://sancdb.rubi.ru.ac.za
NANPDB	Northern African Natural Products Database	www.african-compounds.org/nanpdb
NPACT	Plant-based anticancer compounds	http://crdd.osdd.net/raghava/npact
NPCARE	Natural products for cancer gene regulation	http://silver.sejong.ac.kr/npcare
TIPDB	Traditional Chinese Medicine (focus on Taiwan)	http://cwtung.kmu.edu.tw/tipdb

(Continued)

Table 6.1 (Continued) Most Relevant Databases as References for Natural Product Research

Name	Type	Link
NATURAL PRODUCTS IN PUBCHEM SUBSTANCE DATABASE	Natural product bioactivity	http://ncbi.nlm.nih.gov/pcsubstance
STREPTOMEDB	Natural products produced by streptomycetes	www.pharmaceutical-bioinformatics.de/streptomedb
NUBBE DATABASE	Brazilian natural products	http://nubbe.iq.unesp.br/portal/nubbedb.html
CAROTENOIDS DATABASE	Natural carotenoids biological function	www.carotenoiddb.jp
ANTIBASE	Natural products with antimicrobial activity	http://wwwuser.gwdg.de/~hlaatsc/antibase.htm
DMNP	Marine NPs and their derivatives	http://dmnp.chemnetbase.com
MARINLIT	Marine NPs	http://pubs.rsc.org/marinlit
HERBMEDPRO	American herbal council's database (evidence-based herbal effects)	http://www.herbmed.org/
LONGWOOD HERBAL TASK FORCE	Dietary supplements, herbs, vitamins, and minerals	http://www.longwoodherbal.org/
NATURAL MEDICINES	Dietary supplements, natural medicines, and complementary alternative and integrative therapies	https://naturalmedicines.therapeuticresearch.com/
MSKCC	Herbs database focusing on anticancer activity	https://www.mskcc.org/cancer-care/diagnosis-treatment/symptom-mangement/integrative-medicine/herbs

Source: Boehm, K. et al., *Health Info. Libr. J., 27,* 93–105, 2010; Ya, C. et al., *J. Chem. Inf. Model, 57,* 2099–2111, 2017.

TMCID is a great beginning to make the Chinese data and herbal ingredient more usable for western medicine and drug development, but some details are missing, which could be better developed, for instance to show the number of studies with compounds and effects or to filter the study type.

Another important region of traditional medicinal knowledge is Africa. African biodiversity has huge potential in drug discovery and development; however, it has been limited to random screening of extracts based on ethnobotanical information [47]. AfroDb contains information about African medicinal plants and their effects with the geographical region of plant, and phytochemical data. The database includes maximum 10 tautomers per compound in the dataset with some other chemical properties like the logarithm of the octanol/water partition coefficient representing the lipophilicity factor (log P) and Lipinski violations. Moreover, it contains data from published articles, theses, and textbook chapters as well as unpublished conference presentations from communication with the authors.

6.2.2 Scientific informatics and computational approaches in botanical drug research

It is necessary to use these databases as widely as possible; for example, adapt them to different techniques to increase the efficacy of drug discovery and development. There are many research groups that utilize computer-aided drug design on natural products, including *in silico* screening, while a detailed mechanism of action with the target protein, is known just for a few natural products. These studies could support the cost and time optimization of these research projects and the better understanding of the complexity of natural products. Noteworthy, that the adequate quality of the data is prerequisite as the success of any computational technique is dependent on the quality of the data at the first place [48].

Keum et al. work on *in silico* screening methods to predict interactions between compounds and target proteins of natural products. Their work is based on the theory that the interaction can be predicted by using the structural similarity of compounds and the genomic sequence similarity [49]. There are more methods to identify these similarities, for example, normalized Smith-Waterman scores for sequence similarities between proteins or SIMCOMP, a graph-based method for comparing chemicals structures [40]. In this case, a bipartite local model (BLM) were used, which predicts target proteins of a given protein using the structural similarity of compounds, genomic similarity and information of interactions between compounds and targets. The data of compounds, target proteins, and interactions were collected from the DrugBank database and classified into six groups (G-protein-coupled receptors, enzymes, transporters, receptors and other proteins) [50]. After that, Open Babel fingerprint were used to calculate the compound structure similarity matrices of each type. Genomic sequence similarity matrices of each type were calculated by using Smith-Waterman algorithm and binary interaction matrices of each type are made using information of interactions between compounds and target proteins. Next, bipartite local models are made for predicting interactions between compound and target proteins using these matrices. At last, herb data were taken from databases such as TCMID, TCM-ID, Korean Traditional Knowledge Portal (KTKP) and Japanese Traditional Medicine and Therapeutics (KAMPO) [51].

6.3 Success stories

A few successful applications were collected in this section to highlight the necessity of computational approaches to facilitate natural product discovery workflow. We restrict ourselves to list some representative studies to demonstrate that computer-aided drug design can be and must be adapted for botanical drug research (see Table 6.2).

Table 6.2 Relevant Success Stories That Can Shape the Natural Product Research

Natural Product	Applied Computer-Aided Techniques	Important Results
Licorice (Botanical Drug)	Drug-likeness prediction	It provides that novel in silico strategy for investigation of the botanical drugs containing a huge number of components, which has been demonstrated by the well-studied licorice case.
	Oral bioavailability prediction	
	Blood–brain barrier permeation prediction	
	Target identification and validation	
	Network pharmacology analysis	
Meridianins A–G, isolated from the marine tunicate genus *Aplidium*	Virtual profiling	Meridianins can be classified as kinase inhibitors and used as a starting point to design and develop novel anti-Alzheimer's disease drugs.
	Structure modelling	
	Docking calculations	
	Molecular dynamics simulations	
	Molecular Mechanics/Generalized Born Surface Area (MM/GBSA)	
	Interaction analysis	
	Sequence analysis	
	Selectivity analysis	
	Pharmacokinetic properties	
Streptomyces roseosporus	In silico modeling to identify the corresponding ligands	Better understanding the mechanism of action and more new ligands were identified.
	Homology modeling of a protein	
	Molecular docking	

(Continued)

Table 6.2 (*Continued*) Relevant Success Stories That Can Shape the Natural Product Research

Natural Product	Applied Computer-Aided Techniques	Important Results
Bacillus subtilis, B. amyloliquefaciens, Bacillus sp. KCB14S006	In silico modeling to identify the corresponding ligands Homology modeling of a protein Molecular docking	Better understanding the mechanism of action and more new ligands were identified.
B. amyloliquefaciens, B. subtilis	In silico modeling to identify the corresponding ligands Homology modeling of a protein Molecular docking	Better understanding the mechanism of action and more new ligands were identified.
B. subtilis, B. amyloliquefaciens, B. velezensis	In silico modeling to identify the corresponding ligands Homology modeling of a protein Molecular docking	Better understanding the mechanism of action and more new ligands were identified.
Paenibacillus polymyxa, P. amylolyticus	In silico modeling to identify the corresponding ligands Homology modeling of a protein Molecular docking	Better understanding the mechanism of action and more new ligands were identified.

Sources: Llorach-Pares, L. et al., *Mar. Drugs,* 15, 366, 2017; Jujjavarapu, S.E. and Dhagat, S. *Probiotics Antimicrob. Proteins,* 10, 129–141, 2017; Liu, H. et al., *J. Ethnopharmacol.,* 146, 773–793, 2013.

6.4 Conclusions

The natural-product-inspired design has always played an important role in the drug discovery sciences. The re-emerging interest in natural products as sources for innovative drug discovery may benefit from modern cheminformatics and bioinformatics approaches [55]. While advanced chemical synthesis methods for straightforward modifications of selected moieties of natural products have become generally applied, we believe that a growing interest in computational design methods applied to natural product-derived drug design is inevitable for future success. Several pioneering studies have demonstrated lately, with computational approaches showcased as complementary drivers, that the efficient exploration of botanical drugs for drug discovery can steer drug discovery processes. These powerful tools can save time and money in experiments. Moreover, these methods could point out the best direction to follow for these technologies to maximize the success of natural product prospection, as well as to protect biodiversity [52].

Natural products can in fact be fitted for target repurposing, although they often include undesirable functional groups and substructure moieties. If this detail actually influences natural-product-inspired hit-to-lead progression depends on the particular discovery project. In parallel, the systematic discovery of the protein targets of natural products and their fragments has become tangible due to the availability of predictive computational methods, which will accelerate discovery programs. Importantly, with our increasing knowledge of pharmacologically active metabolites of natural products, computational tools for predicting both the structures and the pharmacokinetic and pharmacodynamics properties of these metabolites will help categorize natural-product-derived lead structure candidates. Given that several of these software tools are openly available to the community or can either be licensed, we foresee increased interest in related paradigms by both academic researchers and pharmaceutical companies. Specifically, we assume that these tools may be deployed to probe traditionally challenging signaling pathways or prototyping ligands for novel disease-relevant drug targets [56]. We still need to explore the domains of application and understand the limitations of natural-product-inspired computational molecular design and target prediction. By maintaining some healthy skepticism, we feel that the time is ripe for (re)discovering and exploiting computational natural product analysis for chemical biology and drug discovery.

References

1. World Health Organisation (WHO). 2013. WHO traditional medicine strategy 2014–2023. *World Health Organization (WHO)* 1–76.
2. Zhang, X. 1998. Regulatory situation of herbal medicines: A worldwide review. *World Health Organization* 1–49. http://apps.who.int/medicinedocs/pdf/whozip57e/whozip57e.pdf
3. European Commission, Herbal Medicinal Products–Major developments. https://ec.europa.eu/health/human-use/herbal-medicines_en.
4. Traditional Chinese Medicine: In Depth, NIH-NCCIH. Traditional Chinese medicine: In depth. https://nccih.nih.gov/health/whatiscam/chinesemed.htm.
5. Ayurvedic Medicine: In Depth, NIH-NCCIH. Ayurvedic medicine: In depth. https://nccih.nih.gov/health/ayurveda/introduction.htm.
6. Veregen® is a registered trademark of Fougera Pharmaceuticals Inc. http://www.veregen.com/.
7. Fulyzaq® is a registered trademark of Napo Pharmaceuticals, Inc. http://www.fulyzaq.com/.
8. FDA approves first anti-diarrheal drug for HIV/AIDS patients [press release], 2012. US Food and Drug Administration, Silver Spring, MD. Available at: www.fda.gov/NewsEvents/Newsroom/PressAnnouncements/ucm333701.htm?source=govdelivery.
9. Duke, J. 2001. Handbook of medicinal herbs. In: *Herbal Reference Library* 677. doi:10.1186/1746-4269-7-30.
10. Ahn, K. 2017. The worldwide trend of using botanical drugs and strategies for developing global drugs. *BMB Reports* 50: 111–16. doi:10.5483/BMBRep.2017.50.3.221.
11. Krause, J., and Tobin, G. 2013. Discovery, development, and regulation of natural products. In: *Using Old Solutions to New Problems-Natural Drug Discovery in the 21st Century* (Ed.) M. Kulka, pp. 3–36. Intech. doi:10.5772/56424.
12. Atanasov, A. G., Waltenberger, B., Pferschy-Wenzig, E. M., Linder, T., Wawrosch, C., Uhrin, P., Temml, V. et al. 2015. Discovery and resupply of pharmacologically active plant-Derived natural products: A review. *Biotechnology Advances* 33: 1582–1614. doi:10.1016/j.biotechadv.2015.08.001.

13. Benzie, I. F. F., and Watchel-Galor, S. 2011. Herbal medicine: An introduction to its history, usage, regulation, current trends, and research needs. *Herbal Medicine: Biomolecular and Clinical Aspects* 464. doi:10.1201/b10787-2.
14. Newman, D. J. and Cragg, G. M. 2014. Natural products as sources of new drugs from 1981 to 2014. *Journal of Natural Products* 79: 629–661. doi:10.1021/acs.jnatprod.5b01055.
15. Atakan, Z. 2012. Cannabis, a complex plant: Different compounds and different effects on individuals. *Therapeutic Advances in Psychopharmacology* 2: 241–254. doi:10.1177/2045125312457586.
16. Brenneisen, R. 2007. Chemistry and analysis of phytocannabinoids and other cannabis constituents. In: *Marijuana and the Cannabinoids* (Ed.) M. A. ElSohly, pp. 17–49. Humana Press, New York. doi:10.1007/978-1-59259-947-9_2.
17. Russo, E. B. 2011. Taming THC: Potential cannabis synergy and phytocannabinoid-Terpenoid entourage effects. *British Journal of Pharmacology* 163: 1344–1364. doi:10.1111/bph.2011.163.
18. Simon, Y. K., Gubili, J., and Cassileth, B. 2008. Evidence-based botanical research: Applications and challenges. *Hematology/Oncology Clinics of North America* 22: 661–670. doi:10.1016/j.hoc.2008.04.007.
19. Boehm, K., Raak, C., Vollmar, H. C., and Ostermann, T. 2010. An overview of 45 published database resources for complementary and alternative medicine. *Health Information and Libraries Journal* 27: 93–105. doi:10.1111/j.1471-1842.2010.00888.x.
20. Harvey, A. L. 2008. Natural products in drug discovery. *Drug Discovery Today* 13: 894–901. doi:10.1016/j.drudis.2008.07.004.
21. Dighe, N. et al. 2015. Herbal database management. *Systematic Reviews in Pharmacy* 1(2): 152. doi:10.4103/0975-8453.75067.
22. Ya, C., De Bruyn Kops, C., and Kirchmair, J. 2017. Data resources for the computer-Guided discovery of bioactive natural products. *Journal of Chemical Information and Modeling* 57: 2099–2111. doi:10.1021/acs.jcim.7b00341.
23. Chinese Pharmacopoeia Commission. 2017. *Pharmacopoeia of the People's Republic of China 2015* (English edition). China Medical Science Press, Beijing, China. http://wp.chp.org.cn.
24. German Pharmacopoeia. 2004–2018. Kooperation Phytopharmaka GbR, Bonn. http://www.koop-phyto.org.
25. Raut, A. A., Chorghade, M. S., and Vaidya A. D. B. 2016. Reverse pharmacology. innovative approaches in drug discovery: Ethnopharmacology. *Systems Biology and Holistic Targeting*. doi:10.1016/B978-0-12-801814-9.00004-0.
26. Dimpfel, W. 2013. Pharmacological classification of herbal extracts by means of comparison to spectral EEG signatures induced by synthetic drugs in the freely moving rat. *Journal of Ethnopharmacology* 149: 583–589. doi:10.1016/j.jep.2013.07.029.
27. Kennedy, D. O. D., and Wightman, E. E. L. 2011. Herbal extracts and phytochemicals: Plant secondary metabolites and the enhancement of human brain function. *Advances in Nutrition* 2: 32–50. doi:10.3945/an.110.000117.32.
28. Liu, Y., and Wang, M. W. 2007. Botanical drugs: Challenges and opportunities. contribution to linnaeus memorial symposium 2007. *Life Sciences* 82: 445–449. doi:10.1016/j.lfs.2007.11.007.
29. Kamboj, A. 2012. Analytical evaluation of herbal drugs. *Drug Discovery Research in Pharmacognosy* 23–60. doi:10.5772/1903.
30. Jiangyong, G., Gui, Y., Chen, L., Yuan, G., Lu, H. Z., and Xu., X. 2013. Use of natural products as chemical library for drug discovery and network pharmacology. *PLoS One* 8: 1–10. doi:10.1371/journal.pone.0062839.
31. Mosihuzzaman, M., and Choudhary, M. I. 2008. Protocols on safety, efficacy, standardization, and documentation of herbal medicine (IUPAC Technical Report). *Pure and Applied Chemistry* 80: 2195–2230. doi:10.1351/pac200880102195.

32. Fullerton, J. N., and Gilroy, D. W. 2016. Resolution of inflammation: A new therapeutic frontier. *Nature Reviews Drug Discovery* 15: 551–567. doi:10.1038/nrd.2016.39.

33. Wells, T. N. C. 2011. Natural products as starting points for future anti-Malarial therapies: Going back to our roots? *Malaria Journal* 10: S3. doi:10.1186/1475-2875-10-S1-S3.

34. Bastin, S. 1999. Herbal medicine 101: The good, the bad & the ugly. Cooperative Extension Service, University of Kentucky, College of Agriculture. Issued 07–98, Revised 08–99; FN SS.084LG.

35. Parveen, A., Parveen, B., Parveen, R., and Ahmad, S. 2015. Challenges and guidelines for clinical trial of herbal drugs. *Journal of Pharmacy and Bioallied Sciences* 7: 329. doi:10.4103/0975-7406.168035.

36. Liu, S. H., Chuang, W. C., Lam, W., Jiang, Z., and Cheng, Y. C. 2015. Safety surveillance of traditional chinese medicine. Current and future. *Drug Safety* 38(2): 117–128. doi:10.1007/s40264-014-0250-z.

37. Barril, X. 2017. Computer-Aided drug design: Time to play with novel chemical matter. *Expert Opinion on Drug Discovery* 12(10): 977–980. doi:10.1080/17460441.2017.1362386.

38. Sarkar, I. N. 2015. Challenges in identification of potential phytotherapies from contemporary biomedical literature. In: *Evidence-Based Validation of Herbal Medicine*, (Ed.) P. K. Mukherjee, pp. 363–371. Elsevier, Amsterdam, the Netherlands. doi:10.1016/B978-0-12-800874-4.00017-9.

39. Harvey, A. L., Edrada-Ebel, R., and Quinn, R. J. 2015. The re-Emergence of natural products for drug discovery in the genomics era. *Nature Reviews Drug Discovery* 14: 111–129. doi:10.1038/nrd4510.

40. Muto, A., and Kanehisa, M. 2009. Structural similarity-based approach to characterize crude drug components. *Workshop on Bioinformatics for Medical and Pharmaceutical Research*, Paris, France. P137:1–2. http://www.jsbi.org/pdfs/journal1/GIW09/Poster/GIW09P137.pdf.

41. Wang, Y., Wang, X., and Cheng, Y. 2006. A computational approach to botanical drug design by modeling quantitative composition-Activity relationship. *Chemical Biology and Drug Design* 68: 166–72. doi:10.1111/j.1747-0285.2006.00431.x.

42. Li, S., Han, Q., Qiao, C., Song, J., Cheng, C., L., and Xu, H. 2008. Chemical markers for the quality control of herbal medicines: An overview. *Chinese Medicine* 3: 7. doi:10.1186/1749-8546-3-7.

43. Yuen, J., Tse, S., and Yung, J. 2011. Traditional chinese herbal medicine–East meets west in validation and therapeutic application. In: *Recent Advances in Theories and Practice of Chinese Medicine* (Ed.) H. Kuang, pp. 239–266. InTech, London, UK. doi:10.5772/26606.

44. Ehrman, T. M., Barlow, D. J., and Hylands, P. J. 2007. Phytochemical databases of Chinese herbal constituents and bioactive plant compounds with known target specificities. *Journal of Chemical Information and Modeling* 47: 254–263. doi:10.1021/ci600288m.

45. DrugBank database. This project is supported by the Canadian Institutes of Health Research, Alberta Innovates—Health Solutions, and by The Metabolomics Innovation Centre (TMIC), a nationally-funded research and core facility that supports a wide range of cutting-edge metabolomic studies. OMx Personal Health Analytics. https://www.drugbank.ca/.

46. OMIM® databases-Online Mendelian Inheritance in Man® are registered trademarks of the Johns Hopkins University. https://mirror.omim.org/.

47. Novick, P. A., Ortiz, O. F., Poelman, J., Abdulhay, A. Y., and Pande, V. S. 2013. Sweetlead: An in silico database of approved drugs, regulated chemicals, and herbal isolates for computer-aided drug discovery. *PLoS One* 8: 1–9. doi:10.1371/journal.pone.0079568.

48. Keum, J., Yoo, S., Lee, D., and Nam, H. 2016. Prediction of compound-target interactions of natural products using large-Scale drug and protein information. *BMC Bioinformatics* 17(Suppl 6): 219. doi:10.1186/s12859-016-1081-y.

49. Lounkine, E., Keiser, M. J., Whitebread, S., Mikhailov, D., Hamon, J., Jenkins, J. L., Lavan, P. et al. 2012. Large-scale prediction and testing of drug activity on side-effect targets. *Nature* 486: 361–67. doi:10.1038/nature11159.

50. Zhang, H. P., Pan, J. B., Zhang, C., Ji, N., Wang, H., and Ji, Z. L. 2014. Network understanding of herb medicine via rapid identification of ingredient-target interactions. *Scientific Reports* 4: 1–8. doi:10.1038/srep03719.

51. Rodrigues, T., Reker, D., Schneider, P., and Schneider, G. 2016. Counting on natural products for drug design. *Nature Chemistry* 8: 531–541. doi:10.1038/nchem.2479.

52. Llorach-Pares, L., Nonell-Canals, A., Sanchez-Martinez, M., and Avila, C. 2017. Computer-Aided drug design applied to marine computer-aided drug design applied to marine. *Marine Drugs* 15: 366. doi:10.3390/md15120366.

53. Jujjavarapu, S. E., and Dhagat, S. 2017. In silico discovery of novel ligands for antimicrobial lipopeptides for computer-aided drug design. *Probiotics and Antimicrobial Proteins* 10(4): 129–141. doi:10.1007/s12602-017-9356-9.

54. Liu, H., Wang, J., Zhou, W., Wang, Y., and Yang, L. 2013. Systems approaches and polypharmacology for drug discovery from herbal medicines: An example using Licorice. *Journal of Ethnopharmacology* 146: 773–793. doi:10.1016/j.jep.2013.02.004.

55. Lahlou, M. 2013. The success of natural products in drug discovery. *Pharmacology & Pharmacy* 4: 17–31. doi:10.4236/pp.2013.43A003.

56. Pal, S. K., and Shukla, Y. 2003. Herbal medicine: Current status and the future. *Asian Pacific Journal of Cancer Prevention* 4(80): 281–288. http://journal.waocp.org/article_24044_ee18e33634f4f05f228690dfaaaec11d.pdf.

7

Botanical Drug Products of Ayurved, Indian System of Medicine in Integration with Other Metallopharmaceuticals, Nano Medical Devices, and Theranostics

Jayant N. Lokhande, Sonali Lokhande, Sameer Mahajan,
Ganesh Shinde, and Yashwant Pathak

Contents

7.1 Botanical drug products with metallopharmaceutics

Most drugs used today are purely organic compounds. Especially after the enormous success of the cisplatin in tumor treatment, interest in metal complexes has grown [1]. The unique physicochemical properties of therapeutic organometallic complexes offer multiple advantages in the discovery and development of new drugs. These metal complexes are amenable to combinatorial synthetic methods and have an immense diversity of structural scaffolds. The most important feature is centres in organometallic complexes that are capable of organizing surrounding atoms to achieve pharmacophore geometries. Additionally, the effects of metals can be highly specific and modulated by recruiting cellular processes that recognize specific types of metal-macromolecule interactions.

Organometallic lead compounds can influence both DNA and RNA with a high degree of regiochemical, sequential, and conformational specificity. Organometallic compounds can target DNA and mRNA expression that can become an attractive preposition with specific cell selectivity. Organometallic compounds are also useful in active site recognition and can act as bifunctional agents as secondary contacts to increase inhibitor affinity. DNA has often been proposed as the target of these organometallic antineoplastic agents; there is a particular emphasis on those that can interact with nucleic acids [2].

Metal complexes can be potent and highly selective ligands of cell surface receptors. Also, studies of toxicity mechanisms may provide insights into potential therapeutic approaches for organometallic complex. Several metals (e.g., vanadium and chromium) appear to have significant effects on complex metabolic diseases (e.g., diabetes) [3].

Some successful case studies of metallopharmaceuticals as per below.

7.1.1 Platinum

Platinum as central atoms containing compounds cisplatin or carboplatin are currently used in cancer chemotherapy. In particular, cisplatin is highly effective in treating ovarian and testicular cancers [4,5]. Carboplatin exhibits a tumor-inhibiting profile identical to that of cisplatin, but with fewer side effects, whereas oxaliplatin is used in a combination therapy against metastatic colorectal cancer. A key factor that might explain the reason that platinum is most useful comes from the ligand-exchange kinetics.

Platinum complexes display, along with other kinds of anticancer drugs, two major drawbacks:

a. severe toxicities (neurotoxicity, nephrotoxicity, etc.)
b. limited applicability to a narrow range of tumors, as several of them exhibit natural or induced resistance [6]

7.1.2 Ruthenium

Ruthenium is classified as the "Platinum group" along with Rhodium, Palladium, Iridium. All these metals often occur together in the same mineral deposits and have closely related physical and chemical properties [6]. Ruthenium compounds exhibit cytotoxicity against cancer cells, and have analogous ligand exchange abilities to platinum complexes with no cross-resistance with cisplatin, and may display reduced toxicity on healthy tissues by using iron transport [5].

The application of Iron in group 8 metal complexes in anticancer drug design is the ferrocenyl derivative of Tamoxifen (ferrocifen) [7].

Ferrocifenes exhibit anticancer activity against hormone dependent and hormone independent breast cancers [8].

Ruthenium complexes demonstrate similar ligand exchange kinetics to those of platinum(II) antitumor drugs with low toxicity. Due to differing ligand geometry between their complexes, ruthenium compounds bind to DNA affecting its conformation differently than cisplatin and its analogues [9].

Ruthenium complexes tend to accumulate preferentially in neoplastic masses in comparison with normal tissue [10]. They probably use transferrin, for its similarities with Iron, to accumulate in the tumor. A Transferrin-Ruthenium complex can be actively transported into tumor tissues that have high Transferrin-receptor densities. Once bound to the Transferrin receptor, the complex liberates Ruthenium that can be easily internalized in the tumor. Next, Ruthenium(III) complexes likely remain in their relatively inactive Ruthenium(III) oxidation state until they reach the tumor site. Ruthenium compounds offer the potential over antitumor Platinum(II) complexes currently used in the clinic of reduced toxicity, a novel mechanism of action, the prospect of non-cross-resistance and a different spectrum of activity [6].

7.1.3 Gallium

Gallium-based anticancer chemotherapeutics are appreciably progressing in clinical studies [11]. Gallium appears to operate through the displacement of metal ions in iron metabolism or bone. In large part, action of gallium complexes seems to be a consequence of the similarity of gallium(III) to iron(III): Gallium interferes with the cellular transport of iron by binding to transferrin, and also interferes with the action of ribonucleotide reductase, which then results in inhibition of DNA synthesis. KP46 is an orally bioavailable gallium complex, which exerts its antitumoral activity via inhibition of ribonucleotide reductase, induction of S phase arrest, and apoptosis [12]. In preclinical models KP46 was proved to be a stronger anticancer agent than gallium nitrate, and it was effective on a model of tumor-associated hypercalcemia. Nominated from a range of gallium complexes for the clinical stage of development, KP46 has finished phase I trials with the outcome of promising tolerability and evidence of clinical activity in renal cell carcinoma [11].

7.1.4 Titanium

Titanocene, metallocene dihalide complexes MX2Cp2 (where M = Ti, V, Mo, Nb etc., X = halide and Cp = η5-cyclopentadienide) TiCl2Cp2 or MTK4 is the most successful anticancer agent as shown in phase I/II clinical trials [13]. Titanocene dichloride had been recognized as an active anticancer drug against breast and gastrointestinal carcinomas. The anticancer activity of TiCl2Cp2 is due to inhibition of collagenase type IV activity, which is involved in regulation of cellular proliferation, protein kinase C, and DNA topoisomerase II activities. Titanium may also replace iron in transferrin and facilitate cellular uptake into tumor cells. The titanocene dichloride is believed to get accumulate via the transferrin–dependent pathways [6].

7.1.5 Gold

Gold compounds due to its capacity to exist in a variety of different coordination states in the biological system are used for treating arthritis and cancer, they have potential for treating AIDS, malaria, and Chagas disease [14] Gold thiolate complexes were found especially effective at slowing the progression of rheumatod arthritis. Sodium Aurothiomalate (myochrysine), Aurothioglucose (solganol), and Aurothiosulfate (sanochrysine) are water-soluble polymeric antiarthritic compounds that are administered to the patient by injection, so-called injectable or parenteral drugs, while Auranofin, which is only slightly soluble in water, is given to the patient orally in capsule form. The compound Aurocyanide, $[Au(CN)_2]^-$, which is a biotransformation product in chrysotherapy, has been found to inhibit proliferation of HIV in a strain of CD_4^+. The compound [bpza] $[AuCl_4]$, where bpza is the diprotonated-chloride form of a bis-pyrazole ligand, inhibits both reverse transcriptase and HIV-1 protease. Since [bpza][$AuCl_4$] is nontoxic to peripheral blood mononuclear cells in the immune system, the compound is thought to have potential as an anti-HIV agent [6].

Auranofin and other gold compounds inhibit the growth of Plasmodium falciparum, a protozoan parasite carried by Anopheles mosquitoes that causes malaria. The researchers suggested that the mechanism by which the gold compounds inhibit the growth of P. falciparum is related to the ability of the complexes to block the function of the enzyme thioredoxin reductase, TrxR [15].

7.1.6 Nickel

Nickel is an essential component in different types of enzymes such as urease, carbon monoxide dehydrogenase, and hydrogenase [16]. The cytotoxicity of the nickel (II) complexes containing 1,2-naphtoquinone-based thiosemicarbazone ligands (NQTS) was tested on MCF7 human breast cancer cell line and compared to free ligand and another naphthoquinone, commercial antitumor drug etoposide. Furthermore, [Ni(II)(3-methoxy-salophene)] overcame vincristine drug resistance in BJAB and Nalm-6 cells [6].

In 2010, new methoxy-substituted nickel(II)(salophene) derivatives are synthesized and their anticancer properties were investigated [17].

7.1.7 Iron

Organometallic compounds like Iron (III)-salophene with selective cytotoxic and antiproliferative properties have also been used in platinum resistant ovarian cancer cells. The low cytotoxicity of ferrocene, coupled with its lipophilicity, electron density, relative thermal and chemical stability, and its electrochemical behavior suggested that this compound could yield interesting results if incorporated into a known drug [8,18].

Brocard and co-workers combined Chloroquine and ferrocene in the same molecule by inserting a ferrocenyl group into the side chain of Chloroquine, producing a hybrid compound called Ferroquine that is more potent than Chloroquine [6].

7.1.8 Manganese

Manganese-based organometallic complexes having Mn-SOD-like activities can be designed as pharmaceutical compositions and dietetic products for use in oxidative stress, including cancer and inflammatory conditions [19]. Mn-containing Super Oxide Dismutase (MnSOD) are universally present and in eukaryotes MnSODs are localised in the mitochondria. The deficiency in Mn-SOD is supposed to have some significance in the development of rheumatoid arthritis. The mitochondrial antioxidant enzyme manganese-containing superoxide dismutase (MnSOD) functions as a tumor suppressor gene and that reconstitution of MnSOD expression in several human cancer cell lines leads to reversion of malignancy.

Mn(II) and Mn(III) macrocycles appear to be particularly promising a manganese(II) complex with bis (cyclohexylpyridine)-substituted macrocyclic ligand has been designed as a functional mimic of SOD, which was reported to have a significant of inflammation and reperfusion injury [20,21]. Nitrogen containing macrocyclic complexes of manganese(II) have shown anti-microbial activity. Mn(III)5,10,15,20-tetrakis(4-benzoic acid)-porphyrin, which can protect against neurodegeneration and is therefore of potential interest for the treatment of brain diseases such as Parkinson and Alzheimer diseases [5].

7.1.9 Cobalt, copper, vanadium, zinc

Aspirin (acetylsalicylic acid) as nonsteroidal antirheumatics (NSAR) inhibits the enzymes in the cyclooxygenase family (COX). A hexacarbonyldicobalt–aspirin complex is shown to inhibit COX activity differently from Aspirin. Besides the role of NSARs in inflammatory processes, they also seem to be involved in tumor growth. Cobalt as metallo-organic component can be added to Aspirin to improve anti-tumor activity. Cobalt–aspirin complexes can exhibit potential cytostatic effect. Co-Aspirin inhibits both cell growth and the formation of small blood vessels in tumors [4]. Cu(II) complexes of NSAIDs with enhanced anti-inflammatory activity and reduced gastrointestinal (GI) toxicity compared with their uncomplexed parent drug, were developed [22].

NSAIDs, remain either largely inadequate and/or are associated with problematic side effects, e.g., renal insufficiency and failure, GI ulceration, bleeding or perforation (NSAID gastropathy), exacerbation of hypertension and congestive heart failure (CHF).

Due to low toxicity of Cu-based nonsteroidal anti-inflammatory drugs (NSAIDs) have led to the development of numerous Cu(II) complexes of NSAIDs with enhanced anti-inflammatory activity [23].

Cu(II) complexes of NSAIDs such as Piroxicam and Meloxicam can be used as chemopreventive and chemosuppressive agents on various cancer cell lines by inhibiting both at the protein level and/or at the transcription level [24].

Vanadium and zinc can exert insulin-mimetic effects in in vitro and in vivo systems [22]. Both vanadyl and zinc complexes enhanced glucose uptake into the adipocytes without the addition of any hormones. Because of this insulin-mimetic activity, oxovanadium(IV) (vanadyl) and zinc(II) (zinc) complexes are proposed to be potent antidiabetic agents for both type 1 and type 2 Diabetes Mellitus therapy. New types of insulin-mimetic vanadyl and zinc complexes such as Bis(allixinato)oxovanadium(IV), [VO(alx)2], bis(allixinato)zinc(II) [Zn(alx)2], bis(thioallixin-N-methyl)zinc(II) [Zn(tanm)2] were developed as the potent activators of the insulin-signaling pathway. The common mechanism of action of VO(alx)2, Zn(alx)2, and Zn(tanm)2 is their effect on the insulin-signaling pathway; this in turn regulates gene transcription and suppresses lipolysis signaling [6].

Combination of metallopharmaceuticals as ligand and or scaffold with Heterogenous Polymolecular Botanical Drug Candidates can be useful in terms of reducing their side effects and or enhance therapeutic efficacy as shown in Table 7.1. Botanical Drug Products have multiple actives so it can be worked on multiple drug receptors. This feature can be very well applied in targeting various disease metabolic pathways in single shot.

7.2 Nanomaterial containing medical devices with botanical drug products

Nanomaterials are fabricated at a nanoscale, and at nanoscale they exhibit nanostructure-dependent properties, e.g., chemical, mechanical, electrical, biological, optical, or magnetic properties that can be used to restructure manufacturing, energy production, and a host of other fields including medicine [25]. Nanomaterials have increased reactivity due to increased ratio of an object's surface area to its volume, as it gets smaller.

Table 7.1 Botanical Drug Products from Ayurved, Indian System of Medicine and Metallopharmaceuticals in Integration

Metallopharmaceuticals	Conjugated with Botanical Drug Product derived from	Efficacy Enhancement in
Platinum – Cisplatin, Carboplatin	Emblica officinalis, Terminalia chebula, Terminalia belerica	Colon Cancer Breast Cancer And reduction in Neurotoxicity Nephrotoxicity Nausea & Emesis
Ruthenium	Centella asiatica	Colon Cancer and Memory Impairment
Gallium	Tribulus terrestris	Renal Cell Carcinoma and reduction in side effects like, Neutropenia Anemia, Stomatitis Conjunctivitis, Dizziness & Headache
Titanium	Aegle marmelos	Gastrointestinal Carcinoma Breast Cancer and reduction in Nephrotoxicity Hepatotoxicity
Gold	Tinospora cordifolia, Bacopa monierri	Inflammation Auto Immune Disorders Brain Tumors Infectious diseases Brain Degenerative Diseases
Nickel	Azadirachta indica	Lymphomas Breast Cancer
Iron	Sesamum indicum	Ovarian Cancer Cerebral Malaria
Manganese	Moringa oleifera Piper nigrum Piper longum Zingiber officinalis	Inflammation Antioxidant Coagulation Disorders Endothelial Disorders Myocardial Ischemia
Zinc	Bauhinia variegata Gymnema sylvestrae	Type I & II diabetes Lymphosarcoma
Copper	Pluchea lanceolata Eclipta alba	Hepatic cancer
Cobalt	Inula racemosa	Inflammation & Pain

The application of nanomaterials in medical devices is a promising area, and numerous innovative medical devises are being currently used. The most important application types of nanomaterials in medical devises are:

- Nanocoating—that increases biocompatibility and thus improve integration with the surrounding tissues of a variety of medical implants used, for example, in cardiology (stent coating), orthopaedics (coating on joint replacement implants), and dentistry (dental implants). In addition, antimicrobial properties of nanomaterials are used in coatings and also in wound care and medical textiles.

- Nano Implants—mimicking of naturally occurring structures to improve biological, physical, and mechanical characteristics of nano implants.
- Nano Devices for Cardiology & Neurology—by utilizing electrical and magnetic properties of nanomaterials on the nanoscale level, e.g., in cardiac arrhythmia, Parkinson disease [26].

SCF (Supercritical fluid-based) manufacturing techniques are applied for the production of nanoparticles, nanofibers, nanowires, nanotubes, nanofilms, and nanostructured materials that has potentially better performances [27].

Some of Successful Examples of Nano Medical Devices in various Therapeutic Sectors as per below.

7.2.1 Cardiology

- Japan Stent Technology Co Ltd (Japan) Carbofilm™ (coating thickness ≤500 nm), iCarbofilm™ (also known as Bio Inducer Surface coating, <300 nm) or the diamond-like carbon coating (average thickness 35 nm) are used to bring similarity with biological tissues to improve biocompatibility and haemocompatibility.
- Ventracor Ltd, Australia The VentrAssist™ is a magnetically elevated third-generation pump with a diamond-like carbon coating on the blood-contacting surfaces that minimizes thrombosis.
- A modified coating technology is the Dylyn™ technology, which is a diamond-like nanocomposite coating containing Si:O.
- The Inert Carbon Technology (ENDOCOR GmbH, Germany) is a surface modification created by high-speed bombardment of carbon ions under vacuum conditions onto the stent's surface.
- CeloNova Biosciences, Inc (USA) Polyzene®-F surface technology is an advanced surface modification of stents that is thrombo-resistant, promotes rapid endothelialisation.
- Eucatech AG (Germany) The Camouflage® nanocoating (100 nm) is nothing but a modified heparine.
- Hexacath (France) is using coat titanium-nitride-oxide on the surface of the stent, based on a plasma technology using the nanosynthesis of a prespecified gas mixture of nitrogen and oxygen and metal.
- Vascular Concepts Ltd (UK) The ProPass™ stent is a bare metal coronary stent system with a platinum coating (250 nm). The platinum coating ensures microsmooth surface architecture.
- Eurocor GmbH, Germany The DIOR® coronary balloon dilatation catheter is coated with a nanoporous matrix consisting of shellac, a natural resin, and paclitaxel. Drug-coated balloons have already proven effective in clinical trials for the treatment of in-stent restenosis [26].

7.2.2 Neurology

- Second Sight Medical Products Inc (USA) Argus® II Retinal Prosthesis System, 60-electrode (200 μm in diameter) array for visual perceptions by electrically stimulating surviving neurons in patients suffering from retinal degradation (retinitis pigmentosa) [28].
- Bionic Vision Australia, a consortium of universities, research institutes and a hospital, is simultaneously developing three different bionic eye devices for subjects with retinitis pigmentosa. The Early prototype (24 electrodes) and the Wide-View device (98 electrodes) are inserted in the suprachoroidal space, whereas the High-Acuity device (256 electrodes) is an epiretinal visual prosthesis for retinal degeneration [29].
- Nano Retina Inc (Israel), a joint venture of Rainbow Medical Ltd (Israel) and Zyvex Labs LLC (USA), is developing the Bio-Retina. This device incorporates various nano-sized components in one implant.
- Retina Implant AG (Germany) The Alpha IMS subretinal visual prosthesis is a microphotodiode-array ("microchip") with 1500 electrodes (50 × 50 μm each) implanted beneath the retina, specifically in the macular region to recognize faces and read large letters and signs [30].
- Pixium Vision SA (France) epiretinal implant IRIS® Vision Restoration System (clinical trial) and the subretinal implant PRIMA Vision Restoration System (preclinical phase) are using microphotodiodes [31].

7.2.3 Oncology

- FREND™ PSA Plus, a lab-on-a-chip cartridge to conduct quantitative immunoassays for measuring the levels of prostate specific antigen in patients with prostate cancer [32].
- CellCollector® and the CELLSEARCH® system are diagnostic tools for detecting and isolating circulating tumor cells (CTCs) from a patient's blood.
- NanoTherm®, a nanoparticle with a superparamagnetic iron oxide (SPIO) core and an aminosilane coating. NanoTherm® nanoparticles are used to generate hyperthermia, sensitises tumor cells to chemo- or radiotherapy. A high temperature increase, referred to as thermal ablation, damages cell structures, resulting in cancer cell destruction [33].
- Sienna+® is an SPIO nanoparticle tracer that is injected into the breast, after which it collects in the lymph nodes draining the breast. A magnetometer (or magnetic sensor device) is used to detect the tracer and guide the surgeon to breast-draining lymph nodes for biopsy [34].
- Aurolase® therapy uses nanoshells, consisting of a silica core and a gold shell, to convert near infrared light into heat for the thermal ablation of solid tumours [35].

- Nano X Ray NBTXR3 nanoparticles are composed of an inorganic core of crystallized hafnium oxide, specifically designed to enhance the effects of radiotherapy [36].

7.2.4 Orthopedics

- Finceramica MaioRegen® Cartilage scaffold is composed of type I collagen and nanostructured hydroxyapatite and is used for the treatment of osteochondral defects in the knee [37].
- University of Aarhus BoneMaster® hydroxyapatite coating with BoneMaster on the fixation of the hip prosthesis in primary hip alloplasty [38].
- Medtronic The TiMesh® system is a line of plates, screws, and meshes for use in reconstructive neurosurgical procedures, such as cranial flap fixation [39].
- Zimmer Biomet's Trabecular Metal Material is a unique, highly porous biomaterial made from elemental tantalum with structural, functional, and physiological properties similar to that of bone. The material features a 100% open, engineered, and interconnected pore structure to support bony in-growth and vascularization [40].

7.2.5 Surgery

- Sandvik Materials Technology Sandvik Bioline 1RK91, is a precipitation hardening stainless steel having very high strength with good ductility [41].
- GFD Gesellschaft für Diamantprodukte mbH (Ulm, Germany) Diamaze PSD (Plasma Sharpened Diamond) combine the unmatched hardness of diamond with the benefits of an extremely sharp cutting edge [42].
- 3-D Matrix Medical Technology, Inc. (USA) is currently preparing investigational studies for FDA approval of PuraMatrix® as a surgical haemostat in the US. For this, self-assembling peptides that form three-dimensional nanofiber scaffolds are used [43].
- PuraStat® Digmed Health Care is a viscous solution of synthetic peptides. Upon contact with blood, the peptide solution instantaneously self-assembles to form a soft gel. This matrix provides a physical barrier to stop bleeding in a variety of surgical indications [44].
- 3-D Matrix Medical Technology PuraMatrix® gels on blood contact and stops the bleeding via mechanical blocking of the bleeding site for diffuse bleeds while allowing a full, transparent view of the target area [45].

- Genadyne Nanogen Aktiv is a Bio-Cellulose nanostructured matrix made from organic nanofiber. Totally natural fiber based on polysaccharides composed of hemicelluloses protein.
- Nanogen Aktigel is a gel created from the same plant extract as the Nanogen Aktiv. Crucial Healing agents are suspended in a gel for quick absorption into the wound bed [46,47].
- Axcelon Dermacare Inc. Nanoderm™, a new bioengineered product, is a nanofiber-based film of pure cellulose. Its average thickness is 0.05 mm, and its selective permeability allows for the passage of water vapor but inhibits the passage of water and microorganisms [48].
- Organogenesis Inc Dermagraft® has a polygalactic or polyglycolic acid mesh seeded with neonatal fibroblasts to produce 3 D nano structure [49].

Various Botanical Drug Products can be developed through the literary and empirical evidence and experience of Ayurved, Indian System of Medicine that can help improve nanomedical devices structure and functioning. BDPs can be nanostructured through various integrated manufacturing methodologies. Once such BDP's nanoforms are conjugated with current and or futuristic nanomedical devices its usability can be increased multifold as shown in Table 7.2

Table 7.2 Nanomedical Devices Improvements with Botanical Drug Products from Ayurved, Indian System of Medicine

Nano Medical Devices	Conjugated with Botanical Drug Product derived from	Efficacy Enhancement
Cardiology – Medicated Stents Function	Nanocomposites of Terminalia arjuna Boerhavia diffusa Nelumbo nucifera inospora cordifoliaT	To improve anti thrombosis capacity To reduce restenosis
Neurology – Retinal Cells Improvement	Nanocomposites of Emblica officinalis, Terminalia chebula, Terminalia belerica	To improve Retinal Cells Regeneration
Oncology – Thermogenesis in Tumor	Nanocomposites of, Plumbago zeylanica Mucuna premnata	To enhance Tumor Tissue Thermogenesis and destruction
Orthopedics – Fibrosis around Prosthesis Parts	Nano Composites of Leptadenia reticulata Symplocos racemosa	To improve callus generation
Surgery – Anti Bleeding Gels & Suture Threads	Nano Composites of Shalmalia malarbica	To reduce bleeding and improve tissue regenesis.

7.3 Botanical drug products with theranostics

Theranostics is an emerging field of medicine that combines specific targeted therapy based on specific targeted diagnostic tests to offer personalized treatments for complex disorders. This science is a classic integration of nanoscience with diagnostic and therapeutic agents are conjugated into a single thernostic agent, to be used for diagnosis, drug delivery, and treatment response monitoring.

PharmaNetics was the company that conceived the term "theranostics," defining the emerging field of medicine that enables physicians to monitor the effect of antithrombotic agents in patients being treated for angina, myocardial infarction, stroke, and pulmonary and arterial emboli [50].

Theranostics uses specific biological pathways in the human body to acquire diagnostic images and also to deliver a therapeutic dose of radiation to the patient. A specific diagnostic test shows a particular molecular target on a tumor, allowing a therapy agent to specifically target that receptor on the tumor, rather than more broadly the disease and location it presents. This contemporary form of treatment moves away from the one-medicine-fits-all and trial-and-error medicine approach, to offering the right treatment, for the right patient, at the right time, with the right dose, providing a more targeted, efficient pharmacotherapy in the form of theranostics [51].

Important features of Theranostics:

- It can predict risks of disease, diagnose disease, stratify patients, and monitor therapeutic response real time.
- Clinicians can make better-informed decisions on timing, quantity, type of drugs, and choice of treatment procedure based on the relevant information.
- Various protein biomarkers and their respective tests can be developed to predict and monitor drug response.
- Drug side effects can be controlled and monitored.

Some Examples of Theranostics:

- Radioactive iodine (I-131) or Iodine-131 Therapy for the diagnosis and treatment of thyroid cancer.
- In September 25, 1998 FDA granted simultaneous approval for both Genentech's Herceptin® for the treatment of Stage IV breast cancer and Dako's HercepTest® for diagnosis of Her2 overexpression [52].
- Camptosar® (irinotecan) and Invader® UGT1A1 Molecular Assay for colon cancer [53].
- Gleevec® (imatinib mesylate) and BCR-ABL LDT and DAKO C-KIT PharmDx® for gastrointestinal stromal tumors (GIST) [54].
- Purinethol® (mercaptopurine) and laboratory-developed thiopurine methyltransferase test for acute lymphatic leukemia [55].

- Radiolabeled ERBITUX® (cetuximab) for head and neck cancer [56].
- The PreVu POC Test can measure Skin Cholesterol as an important new biomarker for Coronary Artery Disease [57].
- Mylotarg®; Pfizer/Wyeth-Ayerst Laboratories, in May 2000, U.S. Food and Drug Administration (FDA) approved gemtuzumab ozogamicin, the first monoclonal antibody drug conjugate or ADC linking calicheamicin to a monoclonal antibody in for treatment of patients with relapsed CD33+ acute myeloid leukemia (AML) [58].
- Inotuzumab ozogamicin, (Besponsa® (Pfizer/Wyeth) also known as CMC-544 is an antibody-drug conjugate was approved by the U.S. Food and Drug Administration in August 2017, is directed against the CD22+ antigen present on B cells in all patients of B-cell malignancies [59].
- Theranostics Australia is currently offering Lutetium Octreotate Therapy for somatostatin positive tumors, Lutetium PSMA Therapy for metastatic or treatment resistant prostate cancer, Yttrium-90 SIRT Therapy for liver cancer, Iodine-131 Therapy for thyrotoxicosis and thyroid cancer, Radium-223 Therapy for metastatic prostate cancer to bones and Yttrium-90 Radiosynovectomy Therapy for inflammatory synovitis of joints [60].

In the context of nanotechnology, it has and can offer lot of contributions in developing such products.

- Iron oxide nanoparticle-(IONP) based theranostic agents—The superior magnetic properties of IONPs, along with their inherent biocompatibility and inexpensiveness, have made IONPs a material of choice in many bioapplications, such as contrast probes for magnetic resonance imaging (MRI). The magnetic properties of IONPs allow them to accumulate upon the summons of an external magnetic field. IONP can itself play an imaging/therapy dual role, due to its potential in hyperthermia. The underlying mechanism is that IONPs can act as antennae in an external alternating magnetic field (AMF) to convert electromagnetic energy into heat. This feature holds promise in tumor therapy for tumor cells that are more susceptible to elevated temperature than normal cells [61]. IONPs with appropriate coatings can be easily coupled with drug molecules.

  For example, Methotrexate (MTX), an anti-cancer drug, onto an aminated IONP surface. Paclitaxel (PTX) to IONP surfaces through a phosphodiester moiety at the (C-2′)-OH position. Drug molecules can also be co-capsulated with IONPs into polymeric matrices.

  Doxorubicin (DOX) and PTX, along with oleic acid coated IONPs, Doxorubicin into anti-biofouling polymer coated IONPs [62–64].
- Gold Nanoparticles based theranostic agents—Gold nanoparticles (Au NPs) possess many unique features and have been investigated in a variety of imaging related arenas, such as in computed tomography (CT), photoacoustics, and surface-enhanced Raman spectroscopy (SERS).

For example, Paclitaxel was modified at its C-7 position and covalently coupled to 4-mercaptophenol modified Au nanoparticles. The resulting conjugates demonstrated better therapeutic effects than MTX alone both *in vitro* and *in vivo*, which was likely due to the "concentrated effect" and an improved pharmacokinetics of the conjugates. Similarly, protein-based pharmaceutics have been loaded onto Au NPs. Tumor necrosis factor (TNF), for instance, was coupled to PEGylated Au nanoparticles, and the resulting conjugates showed better therapeutic efficacy and less toxicity than native TNF [61].

- Carbon nanotube based theranostic agents—Carbon nanotubes (CNTs) have found potential applications in Raman and photoacoustic imaging and have been studied as drug carriers by a number of research groups. CNTs have a graphite-like structure, which is inert and inhibitive to most conjugation chemistry. The molecular imaging with Single Wall NTs and evaluated the combined Gd3+-functionalized SWCNTs when applied to MRI, and high resolution and good tissue penetration were achieved [65].

  Combination of radioisotopes labeled SWCNTs with radionuclide-based imaging techniques (PET and SPECT) can improve the tissue penetration, sensitivity, and medium resolution. There are many characteristic protein biomarkers which often are overexpressed in cancer cells, and they provide an opening gate for early diagnosis, prognosis, maintaining surveillance following curative surgery, monitoring therapy in advanced disease, and predicting therapeutic response. Many important tumor markers have been extensively applied and used in the diagnosis of hepatocellular carcinoma, colorectal cancer, pancreatic cancer, prostate cancers, epithelial ovarian tumor such as carbohydrate antigen 19-9 (CA19-9), alpha-fetoprotein (AFP), carcinoembryonic antigen (CEA), carcinoma antigen 125 (CA125), human chorionic gonadotropin (hCG), and prostate-specific antigen (PSA) [66].

- Silica nanoparticle based theranostic agents—Generally, silica nanoparticles themselves do not have characteristics for imaging. Instead, they afford an excellent platform that allows facile loading of a broad range of imaging and therapeutic functions, making them a good candidate for theranostic purposes. Drug molecules can be easily introduced into silica nanoparticles during particle formation. For example, 2-devinyl-2-(1 hexyloxyethyl)pyropheophorbide (HPPH), a hydrophobic photosensitizing anticancer drug, into silica matrices [61,67].

The integration of Metallopharmaceuticals with Botanical Drug Product using Nanotechnology can be very well used as Theranostics.

There are more than 5,000 different types of herbomineral nanocomposites biomedicines that are mentioned throughout Ayurveda texts for various types of disorders. These biomedicines are being used safely for thousands of years and can be standardized by using few of modern techniques and developed as regulated theranostic agents.

Conjugated Theranostics can have three components, nanocomposites of selected metal, Botanical Drug Product from selected herb with natural or synthetic linking compound as a biodirecting carrier, and Radioisotope agent. Botanical Drug Product and linking compounds can act as bifunctional Carrier utilized for targeting the conjugate to a biological tissue or organ for a specific nanometal. When the purpose of the procedure is diagnostic, images depicting in vivo distribution of the radioisotope or paramagnetic metal can be made by various means. The distribution and corresponding relative intensity of the detected radioisotope or paramagnetic metal not only indicates the space occupied by the targeted tissue, but may also indicate a presence of receptors, antigens, aberrations, pathological conditions, and/or the like. When the purpose of the procedure is therapeutic, the agent typically contains a radioisotope, and the radioactive agent delivers a dose of radiation to the local site. Such Coordination Complex Target metallopharmaceuticals can be useful in the diagnosis of disease by magnetic resonance imaging or scintigraphy, or in the treatment of disease by systemic radiotherapy.

Typical bio-directing carriers can include hormones, amino acids, peptides, peptidomïmetics, proteins, nucleosides, nucleotides, nucleic acids, enzymes, carbohydrates, glycomimetics, lipids, albumins, mono- and polyclonal antibodies, receptors, inclusion compounds such as cyclodextrins, and receptor binding molecules. Few of synthetic polymers like polyaminoacids, polyols, polyamines, polyacids, oligonucleotides, aborols, dendrimers, and aptamers can be used.

Any radioactive metal ion or paramagnetic metal ion capable of producing a diagnostic result or therapeutic response in a human or animal body or in an in vitro diagnostic assay may be used. The selection of an appropriate metal based on the intended purpose is known by those skilled in the art. In some embodiments, the metal may be selected from the group consisting of Lu, Lu-177, Y, Y-90, In, In-1 1 1, Tc, Tc = O, Tc-99m, Tc-99m = O, Re, Re-186, Re-188, Re = O, Re-186 = O, Re-188 = O, Ga$_5$ Ga-67, Ga-68, Cu, Cu-62, Cu-64, Cu-67, Gd, Gd-153, Dy, Dy-165, Dy-166, Ho, Ho-166, Eu, Eu-169, Sm, Sm-153, Pd, Pd-103, Pm, Pm-149, Tm, Tm-170, Bi, Bi-212, As and As-21 [68,69].

Case study done on Mukta shouktic bhasma (MSB) that is a traditional Ayurved medicinal preparation can be portrayed as Theranostic agent with BDP of Aloe vera.

Mukta shouktic bhasma (MSB) is a calcium-containing bhasma. This biomedicine is synthesized through special calcination of the mother of pearl. MSB is used as an antacid, anti-pyretic, and as a source of calcium. It is also used in tuberculosis, cough, asthma, dysmenorrhea, arthritis, rheumatism, and conjunctivitis [70].

This biomedicine is synthesized through special calcination of mother of pearl as mentioned in the classical Ayurvedic text.

The process of synthesis of *bhasma* is divided broadly into three stages:

- Cleaning (*shodhana*): The mother of pearl fragments were gently crushed to smaller fractions of <10 mm using an agate mortar and pestle. Pieces of mother of pearl were first cleaned with hot water to remove dirt material. The mother of pearl fragments were then immersed in lemon juice (*nimbu swarus*) and boiled for 90 min in a specially prepared hanging sealed earthen pot (*dola yantra*). This process is known as boiling (*swedana*). The solution was filtered off to get the cleaned mother of pearl fragments (*shodhit mukta shoutik*), which were subjected to first calcination. For calcination the cleaned mother of pearl fragments were placed in sealed earthen pot (*sarava samputta*) and subjected to ignition in a traditional furnace (*gaja-puta*) as described in Ayurvedic literature to obtain an intermediate. The stable intermediate can be stored in sealed earthen pot until further use.
- Trituration (*bhavana*): The intermediate obtained after the first calcination was then treated with *Aloe vera* gel and triturated using an automated mortar and pestle at 1000 rpm. The total time of trituration was 8 hours. The mixture was pressed in the form of cakes (*chakrikas*) and dried in the shade for 48 hours. These dried cakes were immediately subjected to further processing.
- Calcination (*marana*): The cakes were calcinated to obtain the intermediate. The procedure was repeated two more times with *bhavana* until the sample showed a positive response to all the traditional tests.

Physical evaluation revealed that MSB is a fine grayish-white powder having a poor flow property with narrow particle size distribution of 1.22–22.52 μm having a mean particle size of 10.20 ± 0.45 μm. A clearly identifiable fraction of MSB particles was below 50 nanometer [70,71].

With consideration of Therapeutic Properties of this medicine in conjugating,

- Radionuclides such as yttrium-90 citrate, rhenium-186 sulphide and erbium-169 citrate
- Nano composites of Aloe vera as BDP
- Muktashukti Bhasma
- TNF inhibitor can be used as coordination complex in Rhuematoid Arthirits. Two types of Coordination Complexes, as a diagnostic with Radionuclide and another as therapeutic without Radionuclide can be employed.

By considering various other therapeutic properties of Mukta Shukti Bhasma other Radionuclides and BDPS and linking compounds can be intermixed to explore avenues in ophthalmic disorders, multi-drug resistant tuberculosis, colon cancers, and ovarian cancers.

7.4 Conclusion

Botanical Drug Products, due to its Polymolecular behavior and having innate capacity to target multireceptors in one-go, can be very well coalesced with Metallopharmaceuticals, Nanomedical Devices, and Theranostics. Such new therapeutic and diagnostic products can open novel avenues in many unmet discovery areas of complex disorders where in single molecule-single receptor approach seems to be failing.

References

1. Allardyce, C. S., Dyson, P. J. (2006) Medicinal properties of organometallic compounds, In: *Bioorganometallic Chemistry (Topics in Organometallic Chemistry), Gerard Simonneaux*, Vol. 17, pp. 177–210, Springer-Verlag, Berlin, Germany.
2. Clarke, M. J. (2003) Ruthenium metallopharmaceuticals, *Coordination Chemistry Reviews*, 236(1–2), 209–233.
3. *Database on Metallopharmaceuticals Prepared for DSIR*, New Delhi, India, URDIP March 2007.
4. Ott, I., Gust, R. (2007) Non platinum metal complexes as anti-cancer drugs, *Archive der Pharmazie*, 340(3), 117–126.
5. Meng, X., Leyva, M. L., Jenny, M., Gross, I., Benosman, S., Fricker, B., Harlepp, S. et al. (2009) A ruthenium-containing organometallic compound reduces tumor growth through induction of the endoplasmic reticulum stress gene CHOP, *Cancer Research*, 69(13), 5458–5466, 1538–7445.
6. Beril Anilanmert. (2012) Therapeutic organometallic compounds, *Pharmacology*, Dr. Luca Gallelli (Ed.), InTech.
7. Jaouen, G., Top, S., Vessieres, A. (2006) Chap 3, Organometallics targeted to specific biological sites: The development of new therapies, In: *Bioorganometallics: Biomolecules, Labeling, Medicine*, G. Jaouen (Ed.), pp. 65–95, Wiley-VCH, Weinheim, Germany.
8. Rafique, S., Idrees, M., Nasim, A., Akbar, H., Athar, A. (2010) Transition metal complexes as potential therapeutic. *Agents Biotechnology and Molecular Biology Reviews*, 5(2), 38–45, 1538–2273.
9. Brabec, V., Nováková, O. (2006) DNA binding mode of ruthenium complexes and relationship to tumor cell toxicity. *Drug Resistance Updates*, 9(3), 111–22, 1368–7646.
10. Rademaker-Lakhai, J. M., van den Bongard, D., Pluim, D., Beijnen, J. M., Schellens, J. H. M. (2004) A phase I and pharmacological study with imidazolium-trans-DMSO-imidazole-tetrachlororuthenate, a novel ruthenium anticancer agent. *Clinical Cancer Research*, 10(11), 3717–3727, 1557–3265.
11. Timerbaev, A. R. (2009) Advances in developing tris(8-quinolinolato)gallium(III) as ananticancer drug: Critical appraisal and prospects. *Metallomics*, 1(3), 193–198, 1756–5901.
12. Dittrich, C., Hochhaus, A., Schaad, S., Salama, C., Jaehde, U., Jakupec, M. A., Hauses, M., Gneist, M., Keppler, B. K. (2005) Phase I and pharmacokinetic study of the oral tris-(8-quinolinolato) gallium(III) complex (FFC11, KP46) in patients with solid tumors: A study of the CESAR Central European Society for Anticancer Drug Research: EWIV. *Journal of Clinical Oncology*; *Proceedings of ASCO Annual Meeting*. (June 2005), Vol 23, (16S), Part I of II, June 1 Supplement, pp. 3205, 0732-183X.
13. Bharti, S. K., Singh, S. K. (2009) Recent developments in the field of anticancer metallopharmaceuticals, *International Journal of PharmTech Research*, 1(4), (October–December 2009), pp. 1406-1420, 0974-4304.

14. Dabrowiak, J. C. (2009) *Metals in Medicine*, pp.191–214, John Wiley and Sons, Ist. Ed., West Sussex, UK,.

15. Sannella, A. R., Casini, A., Gabbiani, C., Messori, L., Bilia, A.R., Vincieri, F.R., Majori, G., Severini, C. (2008) New uses for old drugs. Auranofin, a clinically established antiarthritic metallodrug, exhibits potent antimalarial effects in vitro: Mechanistic and pharmacological implications. *FEBS Letters*, 582(6), 844–847, 0014-5793.

16. Abu-Surrah, A. S., Kettunen, M., Leskelä, M., Al-Abed, Y. (2008) Platinum and palladium complexes bearing new (1R,2R)-(-)-1,2-Diaminocyclohexane (DACH)-based nitrogen ligands: Evaluation of the complexes against L1210 Leukemia. *Zeitschrift für anorganische und allgemeine Chemie*, 634(14), 2655–2658, 1521–3749. doi:10.1002/zaac.200800281.

17. Lee, S., Hille, A., Frias, C., Kater, B., Bonitzki, B., Wolf, S., Scheffler, H., Prokop, A., Gust, R. (2010) [NiII(3-OMe-salophene)]: A potent agent with antitumor activity. *Journal of Medicinal Chemistry*,53(16), 6064–6070, 1520–4804.

18. Blackie, M. A. L., Chibale, K. (2008) Metallocene antimalarials: The continuing quest. *Metal-Based Drugs*, 2008. doi:10.1155/2008/495123, 1687–5486.

19. Maurel, J. C., Cudennec, C. A. (2009) Manganese-based organometallic complexes, pharmaceutical compositions and dietetic products, Assignees: MEDESIS PHARMA S.A., IPC8 Class: AA61K3156FI, USPC Class: 514169, Publication date: 10/01/2009, Patent application number: 20090247492.

20. Guo, Z., Sadler, P. J. (1999) Metals in medicine. *Angewandte Chemistry International*, 38(11), 1512–1531, 1521–3773.

21. Aston, K. Rath, N., Naik, A., Slomczynska, U., Schall, O. F., Riley, D. P. (2001) Computer-aided design (CAD) of Mn(II) complexes: Superoxide dismutase mimetics with catalytic activity exceeding the native enzyme. *Inorganic Chemistry*, 40, 1779–1789, 0020–1669.

22. Hiromura, M., Sakurai, H. (2008) Action mechanism of metallo-allixin complexes as antidiabetic agents. *Pure and Applied Chemistry*, 80(12), 2727–2733.

23. Trinchero, A., Bonora, S., Tinti, A., Fini, G. (2004) Spectroscopic behavior of copper complexes of nonsteroidal anti-inflammatory drugs. *Biopolymers, Special Issue: Selected Papers from the 10th European Conference on the Spectroscopy of Biological Molecules (ECSBM 2003)*, 74(1–2), 120–124, 1097-0102.

24. Roy, S., Banerjee, R., Sarkar, M. (2006) Direct binding of Cu(II)-complexes of oxicam NSAIDs with DNA backbone. *Journal of Inorganic Biochemistry*, 100(8), 1320–1331, 0162-0134.

25. *Nanotechnology: Innovation for Tomorrow's World*. : European Commission, Brussels, Belgium, 2004.

26. RIVM Report 2015-0149 Geertsma, R. E. et al. Nanotechnologies in medical devices, National Institute for Public Health and the Environment, Utrecht, the Netherlands.

27. Fulton, J. L., Deverman, G. S., Yonker, C. R., Grate, J. W., De Young, J., and McClain, J. B. (2003) Thin fluoropolymer films and nanoparticle coatings from the rapid expansion of supercritical carbon dioxide solutions with electrostatic collection. *Polymer*, 44, 3627–32.

28. Humayun, M. S., Weiland, J. D., Fujii, G. Y., Greenberg, R., Williamson, R., Little, J., Mech, B. et al. (2003) Visual perception in a blind subject with a chronic microelectronic retinal prosthesis. *Vision Research*, 43, 2573–2581.

29. Ahnood, A., Escudie, M. C., Cicione, R., Abeyrathne, C. D., Ganesan, K., Fox, K. E., Garrett, D. J. et al. (2015) Ultrananocrystalline diamond-CMOS device integration route for high acuity retinal prostheses. *Biomed Microdevices*, 17, 9952.

30. Stingl, K., Bartz-Schmidt, K. U., Besch, D., Chee, C. K., Cottriall, C. L., Gekeler, F., Groppe, M. et al. (2015) Subretinal visual implant alpha IMS: Clinical trial interim report. *Vision Research*, 111, 149–60.

31. Lorach, H. and Goetz, G. (2015) Photovoltaic restoration of sight with high visual acuity. *Nature Medical*, 21, 476–82.

32. Hermsen, S. A. B., Roszek, B., van Drongelen, A. W., and Geertsma, R. E. (2013) Lab-on-a-chip devices for clinical diagnostics: Measuring into a new dimension. RIVM Report 080116001. National Institute for Public Health and the Environment, Bilthoven, the Netherlands.

33. Gunaratnam, E., James, K., Vasdev, N. (2015) The increasing application of nanotechnology and nanomedicine in urological oncology. *JSM Nanotechnology Nanomedicine*, 3.

34. Foerster, V. (2014) *SentiMag® and Sienna+® for Sentinal Lymph Node Localisation in Breast Cancer*. Health Policy Advisory Committee on Technology (HealthPACT), State of Queensland, Australia.

35. Jain, K. K. (2015) *Nanobiotechnology: Applications, Markets and Companies*. Jain PharmaBiotech, Basel, Switzerland.

36. http://lifesciences.instinctif.com/news/2015/01/nanobiotix-2014-review-2015-anticipated-milestones-and-financial-calendar visited on 8/8/2018.

37. https://jri-ltd.com/product-range/orthobiologics/maioregen/ visited on 8/8/2018.

38. https://clinicaltrials.gov/ct2/show/NCT02311179 visited on 8/8/2018.

39. http://www.medtronic.com/us-en/healthcare-professionals/products/neurological/cranial-repair/timesh-cranial-plating-system.html visited on 8/8/2018.

40. https://www.zimmerbiomet.com/medical-professionals/common/our-science/trabecular-metal-technology.html visited on 8/8/2018.

41. https://www.materials.sandvik/en-us/products/wire/high-performance-materials/sandvik-1rk91/ visited on 8/8/2018.

42. http://en.blades.diamaze-gfd.com visited on 8/8/2018.

43. http://www.puramatrix.com/wp/?page_id=158 visited on 8/8/2018.

44. http://diagmed.healthcare/product/purastat-haemostasis-agent/ visited on 8/8/2018.

45. http://www.puramatrix.com visited on 8/8/2018.

46. https://www.genadyne.com/products/nanogen-aktiv visited on 8/8/2018.

47. https://www.genadyne.com/collections/nanogen/products/nanogen-aktigel?variant=27484953478 visited on 8/8/2018.

48. http://nanoderm.ca visited on 8/8/2018.

49. https://organogenesis.com/products/dermagraft.html visited on 8/8/2018.

50. http://oralcancernews.org/wp/theranostics-guiding-therapy/ visited on 8/8/2018.

51. http://theranostics.com.au/what-is-theranostics/ visited on 8/8/2018.

52. https://www.ddw-online.com/personalised-medicine/p148484-theranostics-an-emerging-tool-in-drug-discovery-and-commercialisation-fall-02.html visited on 8/8/2018.

53. https://www.centerwatch.com/drug-information/fda-approved-drugs/drug/131/campostar visited on 8/8/2018.

54. https://www.us.gleevec.com/#cobrand-1 visited on 8/8/2018.

55. https://www.accessdata.fda.gov/drugsatfda_docs/label/2011/009053s032lbl.pdf visited on 8/8/2018

56. https://www.erbitux.com/?utm_campaign=erbsem&utm_source=ggl&utm_medium=ppc&utm_content=%7Cbr%7Cbr-gen%7C43700023454210182%7Cp&utm_keyword=Erbitux&gclid=CKHw_sqp7dwCFRmSxQIdhqIGAA&gclsrc=ds visited on 8/8/2018

57. http://www.prevu.com/English/hcp/skinchol.html visited on 8/8/2018

58. https://adcreview.com/adc-university/adcs-101/cytotoxic-agents/calicheamicin/ visited on 8/8/2018

59. https://adcreview.com/inotuzumab-ozogamicin-cmc-544-drug-description/ visited on 8/8/2018

60. http://theranostics.com.au/what-is-theranostics/ visited on 8/8/2018

61. Jin, X., Seulki, L., Xiaoyuan, C. (2010) Nanoparticle-based theranostic agents. *Advanced Drug Delivery Reviews*, 62(11), 1064–1079. doi:10.1016/j.addr.2010.07.009.

62. Kohler, N., Sun, C., Wang, J., Zhang, M. Q. (2005) Methotrexate-modified super-paramagnetic nanoparticles and their intracellular uptake into human cancer cells. *Langmuir.* 21, 8858–8864.

63. Yu, M. K., Jeong, Y. Y., Park, J., Park, S., Kim, J. W., Min, J. J., Kim, K., Jon, S. (2008) Drug-loaded superparamagnetic iron oxide nanoparticles for combined cancer imaging and therapy in vivo. *Angewandte Chemie International Edition England,* 47, 5362–5365.

64. Jain, T. K., Richey, J., Strand, M., Leslie-Pelecky, D. L., Flask, C. A., Labhasetwar, V. (2008) Magnetic nanoparticles with dual functional properties: Drug delivery and magnetic resonance imaging. *Biomaterials,* 29, 4012–4021.

65. Hong, H., Gao, T., Cai, W. (2009) Molecular imaging with single-walled carbon nanotubes. *Nano Today.* 9(3), 252–261.

66. Eatemadi, A., Daraee, H., Karimkhanloo, H., Kouhi, M., Zarghami, N., Akbarzadeh, A., Joo, S. W. et al. (2014) Carbon nanotubes: Properties, synthesis, purification, and medical applications. *Nanoscale Research Letters,* 9(1), 393. doi:10.1186/1556-276X-9-393.

67. Roy, I., Ohulchanskyy, T. Y., Pudavar, H. E., Bergey, E. J., Oseroff, A. R., Morgan, J., Dougherty, T. J., Prasad, P. N. (2003) Ceramic-based nanoparticles entrapping water-insoluble photosensitizing anticancer drugs: A novel drug-carrier system for photo-dynamic therapy. *Journal of the American Chemical Society* 125, 7860–7865.

68. U.S. Patent No. 5,435,990 Roberta, C. C., William, A. F., William, F. G., William, J. K., Richard, K. F., Joseph, R. G., Garry, E. K., Kenneth, M., Jaime, S., David, A. W., Sharon, B. Macrocyclic congugates and their use as diagnostic and therapeutic agents 1988-06-24.

69. US20110177004A1 Dennis A. Moore Version of FDG Detectable by Single-Photon Emission Computed Tomography granted on 2014-10-14.

70. Sharma, R. N. 1985. *Ayurveda-sarsangrha* (13th Edn), Shri Baidhyanath Ayurveda Bhavan Ltd. Varanasi, India, pp. 101–102 (in Hindi).

71. Nitin, D., Nidhi, D., Rajendra, S. M., Ajay, K. S., Dinesh, K. J. (2009) Physicochemical and pharmacological assessment of a traditional biomedicine: Mukta shouktic bhasma. *Songklanakarin Journal of Science and Technology* 31(5), 501–510.

8

The Quality, Safety, and Efficacy of Botanical Drugs

Kimberly Palatini and Slavko Komarnytsky

Contents

8.1 Introduction

In the United States, botanical products with health-related claims can be marketed as conventional foods, dietary supplements, or drugs, depending on the specific claim[1] (Table 8.1). These specifications were passed in the Dietary Supplement Health and Education Act (DSHEA) of 1994 and further discussed in the Food and Drug Administration (FDA) industry guidelines of 2004 and 2015.[2] Under the DSHEA, a botanical product is considered a drug only if it bears a disease claim.[3] This is different from the approaches adopted by European and Canadian regulatory authorities, which put botanical remedies in their own category with a unique standard of review. Although manufacturers of botanical products cannot legally make disease claims without the approval of a new drug application (NDA), unsubstantiated medical uses for many botanicals in the US are well known and often promoted in literature and on the Internet. Since many people other than manufacturers can place overt disease claims into the community, it is inevitable that botanical products are often used by self-medicating consumers as drugs without regulatory assurance of efficacy and safety.[1]

Historically, botanicals were well-established medicines listed in the US Pharmacopeia (USP) and prescribed by physicians, but this practice ended with strict drug safety and efficacy guidelines issued by the FDA in 1962. Only a few botanical medicines were grandfathered and allowed to remain on the market as long as their ingredients and labeling remain the same, for example Lydia E. Pinkham's Vegetable Compound first marketed in 1875 and still on the market today. The rapid removal of botanical remedies from the USP pushed many of the remaining products to the over-the-counter (OTC) drug marketplace under Category II (not recognized as safe or effective) or Category III (safety and efficiency not yet determined) status. Once the FDA completes the ongoing OTC drug review, only few botanicals are expected to remain as the OTC active ingredients due to lack of commercial sponsors and required safety and efficiency data. Since the FDA does not consider safety and efficacy of botanical remedies sold in foreign countries as a sufficient evidence for the OTC drug candidacy in the US, the modern botanical industry has therefore chosen to develop and market their products as foods (dietary supplements), some of which could be found on the Generally Recognized as Safe and Effective (GRASE) list of 1968.

In order to differentiate and best evaluate drug products derived from botanical sources and aimed at the diagnosis, cure, mitigation, treatment, or prevention of a disease, the Center for Drug Evaluation and Research (CDER) at the FDA developed specific regulatory guidelines that apply only to botanical products intended to be developed and used as "botanical drugs."[2] This category includes materials derived from plants, algae, macroscopic fungi, and combinations thereof and excludes (i) products that contain animals or animal parts, and/or minerals except when these are a minor component in a

Table 8.1 Common Classes of Botanical Products in the US

Form	Definition	Examples
Food Additives	Intend to become a component of food	
Food colors not GRAS/FDA approval	Federal Food, Drug, and Cosmetic Act (21 U.S.C. §321(s)(1)): *Any substance the intended use of which results or may reasonably be expected to result, directly or indirectly, in its becoming a component or otherwise affecting the characteristics of any food … is capable (alone or through reaction with other substance) of imparting color thereto.*	Annatto extract Beet powder Carrot oil Turmeric
Functional foods[a] Nutraceuticals[a]	Intend to add further nutritional value to a diet	
Foods for special dietary use same as food	Code of Federal Regulations (21 C.F.R. §105.3): *For supplying particular dietary needs which exist by reason of a physical, physiological, pathological or other condition, including but not limited to the conditions of diseases, convalescence, pregnancy, lactation, allergic hypersensitivity to food, underweight, and overweight; … by reason of age; … for supplementing or fortifying the ordinary or usual diet with any vitamin, mineral, or other dietary properly.*	Hypoallergenic foods Infant foods
Dietary supplements same as food structure/function only	Federal Food, Drug, and Cosmetic Act (21 U.S.C. §321(ff)): *A product (other than tobacco) intended to supplement the diet that bears or contains one or more of the following dietary ingredients: a vitamin, a mineral, an herb or other botanical, an amino acid, a dietary substance for use by man to supplement the diet by increasing the total dietary intake, or a concentrate, metabolite, constituent, extract, or combination of any.*	Omega-3 fatty acids Phytosterols

(*Continued*)

Table 8.1 (Continued) Common Classes of Botanical Products in the US

Form	Definition	Examples
Medical foods same as food medical supervision	Intend to diagnose, cure, mitigate, treat, or prevent a disease Orphan Drug Act (21 U.S.C. 360ee(b)(3)): *A food which is formulated to be consumed or administered enterally under the supervision of a physician and which is intended for the specific dietary management of a disease or condition for which distinctive nutritional requirements, based on recognized scientific principles, are established by medical evaluation.*	Axona Lofenalac
Patent medicines[a] grandfathered	As long as ingredients and labeling remain the same	Pinkham's Vegetable Absorbine Jr.
Botanical drugs same as drug	Federal Food, Drug, and Cosmetic Act (21 U.S.C. 355(b)): *A product that is used as a drug and that contains as ingredients vegetable materials, which may include plant materials, algae, macroscopic fungi, or combinations thereof.*	Fulyzaq Veregen
Pharmaceuticals[a]	Intend to diagnose, cure, mitigate, treat, or prevent a disease	
OTC drugs same as drug GRASE (old) NDA (new) FTC regulations	As long as the active ingredients are listed in the OTC drug monograph	Aspirin Echinacea
Prescription drugs IND/NDA FDA approval CMC requirements	Federal Food, Drug, and Cosmetic Act (21 U.S.C. 201(g)(1)(b)): *An article intended for use in the diagnosis, cure, mitigation, treatment, or prevention of disease in man or other animals; and an article (other than food) intended to affect the structure or any function of the body of man or other animals.*	Morphine Taxol

[a] Terms not specifically defined by law

Table 8.2 Comparative Analysis of Botanical Interventions Described in This Study

Form	Plant	Source	Putative Actives	Formulation
Botanical Drug				
Veregen (Sinecatechins)	Common tea *Camellia sinensis*	Leaf extract	Catechins	Ointment, 15% (topical) 112.5 mg
Fulyzaq (Crofelemer)	Dragon's Blood *Croton lechleri*	Latex extract	Proanthocyanidins	Tablet (oral) 125 mg
Nutraceutical				
Estrovera	Siberian rhubarb *Rheum rhaponticum*	Root extract	Hydroxystilbenes	Tablet (oral) 4 mg

traditional botanical preparation; (ii) materials derived from botanical species that are genetically modified with the intention of producing a single molecular entity; (iii) products produced by fermentation of yeast, bacteria, plant cells, or other microscopic organisms, including plants used as substrates, if the objective of the fermentation process is to produce a single molecular entity; and (iv) highly purified substances, either derived from a naturally occurring source (e.g., paclitaxel) or chemically modified (e.g., estrogens synthesized from yam extracts).[2]

These guidelines also consider the complex nature of botanical therapies and facilitate the development of new therapies from botanical sources. This is especially important because a conventional drug approved by the FDA has a single, well-characterized active ingredient. In contrast, botanical drug products often have unique features, complex mixtures, and lack of a distinct active ingredient. They are traditionally delivered in multiple forms, including teas, powders, topical gels, and poultices with variation between individual preparations. It was therefore critical to provide an alternative path for developing these botanical drug products under the strict quality, safety, and efficacy regulations of prescription drugs. Such an integrated approach is best explained and illustrated through the FDA's experience with the first two botanical NDAs (Mytesi® (Fulyzaq®) and Veregen®) when compared to a dietary supplement (Estrovera), that was developed following the similar evidence-based preclinical and clinical programs, but has not been subjected to the FDA review and approval (Table 8.2).

8.2 Quality, safety, and efficacy of conventional drugs

Before administering an investigational drug to human volunteers, a sponsor must file an investigational new drug (IND) application with the FDA. The IND application includes chemical and manufacturing data, animal test results, and incorporates all pharmacology and safety data gathered during the preclinical stage. These typically include an array of genotoxicity, reproductive toxicity, repeat-dose toxicity in two mammalian species (one rodent), and carcinogenicity (unless exempt for some short-term indications) studies. Toxicokinetic studies to support systemic exposure and safety pharmacology using screens for modes/sites of action are also required. The IND also must explain the rationale for testing a new compound in humans, strategies for protection of human volunteers, and a plan for clinical testing. The FDA has 30 days to interfere before the company can proceed with Phase I testing.[4]

Phase I clinical trials include between 20 and 100 individuals, needing both healthy individuals and those with the disease or condition. At this stage, researchers attempt to understand how the drug interacts with the body, effective dosing ranges, and potential side effects are associated with different doses. This data gives early information about how effective the new drug is to determine the most appropriate dose to limit risk and maximize benefits. Seventy percent of drugs that reach this stage progress to Phase II testing.[3,4]

Phase II trials are larger, including several hundred volunteer patients. Despite a greater number of participants, these tests still are not large enough to definitively demonstrate benefits of the drug. Instead, the goal is to gather additional safety data. The new drug can be evaluated for interactions with other medications and on patients with additional complications. Thirty-three percent of drugs that reach this stage progress to Phase III testing.[3,4]

Phase III trials include hundreds and often thousands of patients in a study that lasts between 1 and 4 years. These trials are large enough to show if a drug will be beneficial to a specific population with the targeted disease or condition. Phase III studies provide safety data that had previously gone unrecorded due to short trial times or small populations. More patients and longer prescriptions allow rare or long-term side effects to present.[4] Often, more than one Phase III trial is required before sponsors can progress to submitting a new drug application (NDA). An NDA is the combined clinical, pharmacological and toxicological data gathered during all three phases of the clinical trial. On average, it takes 3 years from the time an NDA is submitted to the time it is reviewed by the FDA. In total, bringing a new chemical entity (NCE) to market takes 10–15 years on average, and nearly a billion dollars.[4]

8.3 Botanical drug approval: totality of evidence approach

The same FDA staff that oversees approvals of conventional drugs also reviews applications for botanical drugs. However, to ensure consistent implementation of botanical drug guidance, CDER/FDA additionally established the Botanical Review Team (BRT) in 2003. The BRT provides scientific expertise on botanical issues to the reviewing staff, guarantees consistent interpretation and implementation of the Botanical Guidance, consolidates experiences in regulatory review of botanical applications, and compiles information on the status of botanical drug submissions for agency management.[2] Additionally, the BRT provides assistance to sponsors of botanical applications in the interpretation of the regulations and their interaction with the FDA.[1] The purpose of the BRT review is to provide historical background of the botanical, to help the clinical review division better understand the product and to search for information that may be relevant to the new use, but not submitted in the application.

Biology of the medicinal plants, pharmacology of the botanical product, and prior human experience with the botanical product are very complex and often contradictory issues. Despite their substantial prior human use, the complexity of these treatments makes conventional clinical trials difficult and expensive. To address these challenges, the FDA has developed the "totality of evidence" approach based on knowledge and expertise acquired from the review of botanical IND and NDA submissions. In addition to the conventional chemistry, manufacturing and controls (CMC) data, this integrated approach considers other evidence including raw material control, clinically relevant bioassays and other non-CMC data deemed necessary by the BRT. The degree of reliance on these other data for controlling consistency of quality depends on the extent to which the botanical mixture can be characterized and quantified. For example, early adoption of clinically relevant bioassays provides a measure of overall potency of the botanical product, while demonstration of clinical dose response minimizes concern over batch-to-batch variability (Table 8.3).

Table 8.3 Quality, Safety, and Efficacy Requirements for Development and Approval of Botanical Drugs

Requirements	Botanical Drugs
Overall	Same as drug[a]
Quality	
Botanical raw material	Identification by trained personnel
	Certificate of authenticity
	List of all growers and/or suppliers
Botanical drug substance	Qualitative and quantitative description
	Chemical constituents may not be always defined
	Active constituents may not be identified
	Name and address of manufacturer
	Description of manufacturing process
	Quality control tests performed
	Chromatogram fingerprint (presence, not amount)
	Biological assay (dose-dependent)
	Description of container/closure system
	Available stability data
	Container label
Botanical drug product	Same as above (but for a finished product), plus
	Placebo and labeling
	Environmental assessment or claim of categorical exclusion

(*Continued*)

Table 8.3 (*Continued*) Quality, Safety, and Efficacy Requirements for Development and Approval of Botanical Drugs

Requirements	Botanical Drugs
Safety	
Prior history of human use	Documented daily human consumption > proposed trial dose[a]
	Documented duration of human use > proposed trial duration[a]
	Equivalency of amount used in raw form > proposed trial dose[a]
Preclinical data	Volume of sales (ex-US market)[a]
	Drug substance, active and known constituents if feasible[a]
	Data from toxicological databases (RTECS, Toxline, TOMES)[a]
	Extensive literature search (Medline)[a]
	Standard nonclinical toxicology tests for Phase III and NDA
	Genotoxicity
	Reproductive toxicity
	Repeat-dose toxicity in two mammalian species (one rodent)
	Carcinogenicity not required for some short-term indications
	Toxicokinetic studies to support systemic exposure
	Safety pharmacology using screens for modes/sites of action
Efficacy	Controlled efficacy trials (Phase I–III)

[a] May have reduced CMC requirements for initial (Phase I and II) clinical studies

8.4 Unique aspects of botanical drug approval process

Capitalizing on the knowledge of botanical products could present an opportunity for a cheaper alternative to the costly process of conventional drug approval. The advantage of botanical drugs lies in their ability to be directly evaluated for clinical efficacy first, rather than being subjected to pre-clinical testing. Once efficacy is proven, these products can be developed either as standardized heterogeneous mixtures or as purified single-chemical drugs. The novelty of the botanical drug approval process is, therefore, in taking advantage of prior history of human use to jump-start the drug development process. For many botanical preparations, extensive human experience can provide some degree of comfort in their safety, but these past human experiences have rarely been documented with rigorous scientific standards in mind. They are often abundant, but mostly anecdotal and of poor quality. It is therefore challenging to determine how these types of human data can substitute for conventional animal toxicity studies in the safety evaluation.

As the traditional uses of many botanical products are largely based in theory and practice of alternative medicine, interpretation of these experiences has been a problem in designing clinical trials. In standard references, the pharmacology of botanical products is typically complicated because of the complexity of the natural mixtures.[2] The products are often indicated to treat a great variety of seemingly unrelated symptoms, without reference to the mechanism of action or the effect on the underlying diseases. Furthermore, in alternative medicine, the definitions of diagnoses, symptoms, and treatment-related adverse events are often vague and difficult to understand or correlate with medical terminology.[1]

Botanical products derived from multiple or even single plants are complex mixtures of numerous chemical entities. Botanical drugs may contain a single part of one plant, multiple parts of the same plant, or different parts from many plants.[5] Regulations require that the contribution of each component of the fixed combination is shown.[2] Although each of the individual plants in combination may have been used widely, either alone or in combination with other plants, reasons for combining many plants in the specific product are often not clear, and there is rarely good evidence of a contribution to effectiveness. Even for extensively studied plants, only a small fraction of the constituents have been isolated and identified. Complete characterization of each individual compound in botanical drugs, even from just a single plant, remains a daunting task, resulting in the chemical composition of botanicals being only partially defined.[1,5] Strength and potency of these vaguely defined products are difficult to determine and quantify, especially over time, adding to the difficulties in CMC controls and clinical pharmacology studies.

The crucial question for approval of botanical drugs is whether the future marketed batches will have the same therapeutic effect as that observed in clinical trials. Batch-to-batch variation is a common problem that can occur for multiple reasons, making the final product differ significantly between producers. Because plant growth and composition can be affected by soil, weather, seasonal variations, geographic location, and other agricultural practices, tight controls must be imposed on when and where plants can be grown.[1] Not only does that control for the presence of the active compounds but also the presence of contamination from trace heavy metals, residual pesticides, and infectious micro-organisms.[5] Additionally, the variation in processing methods and their potential influence on the therapeutic effects further complicate the quality of botanical products. For these reasons, a detailed description of the raw materials and processes, not just of the drug substances and the final drug product, is required for all botanical products.

8.5 Sinecatechins (Veregen): topical formulation of polyphenols extracted from green tea leaves

Veregen, approved in 2006, is the first botanical drug ever approved by the FDA. It is the only botanical drug currently approved for the treatment of external genital and perianal warts (EGWs) in immunocompetent patients 18 years or older. The active ingredients are sinecatechins, extracted from the leaves of *Camellia sinensis (L.)*, and the drug is prescribed as a topical ointment to be applied 3 times per day (Table 8.2). Green tea catechins present in Veregen exhibit specific anti-oxidant, anti-viral, anti-tumor and immunostimulatory activity, which highly contribute to Veregen's efficacy in treatment of EGWs.[6] Each gram of the Veregen ointment contains 150 mg of sinecatechins in a water-free ointment base consisting of isopropyl myristate, white petroleum, cera alba (white wax), propylene glycol, palmitostearate, and oleyl alcohol.[7] Treatment with Veregen should be continued until complete clearance of all warts, but no longer than 16 weeks.[7]

The mode of action of Veregen involved in the clearance of genital and perianal warts is unknown. In vitro, sinecatechins inhibited the growth of activated keratinocytes and demonstrated anti-oxidative activity locally, yet the clinical significance of this finding is yet unknown.[7] An *in vitro* study documented the inhibition of an extensive range of enzymes and kinases, including oxygenases, proteases, and protein kinases, involved in the generation of inflammatory mediators with micromolar concentrations of Veregen.[8]

8.5.1 Standard indication: genital warts

Human papillomavirus (HPV) is a widespread infection and the most commonly diagnosed sexually transmitted infection (STI) in several geographic regions of the world.[9] It is estimated that 6.2 million new infections occur annually in individuals aged 14–44 years.[10] EGWs, also known as venereal warts or condylomata acuminata (CA), are benign epithelial mucosal tumors caused by HPV. HPV strains 6 and 11 are responsible for 90% of EGW.[11] Generally, these lesions are benign in nature, in contrast to other genital HPV infections caused by high-risk HPV strains 16 and 18 that are associated with cervical cancer.[12] The prevalence of EGW is estimated to be about 1% of sexually active individuals in developed countries and is increasing in several regions of the world. However, clinically detectable cases comprise only a small subset of all cases, with 20 million cases between the US and Europe when including subclinical HPV.[9] Persisting genital warts rarely cause severe complaints such as pain, burning, itching, or obstructions, but can be disfiguring and stigmatizing causing fear of cancer or infertility. It is primarily the psychological distress of those affected that represents the main demand for treatment.[13]

Infections of HPV are strictly confined to the epithelium. The virus initially infects basal keratinocytes, most likely by small traumas of the stratified epithelium. Subsequent steps of viral genome amplification, expression of capsid proteins, and assembly of virus particles are closely linked to the differentiation of keratinocytes as they migrate through the squamous epithelium.[14] Formation of mature virus particles in the most superficial epithelial layers does not elicit danger signals, since these cells are destined to die. Furthermore, HPVs are able to inhibit type 1 IFN responses, an anti-viral defense mechanism present in all cells, including keratinocytes. As a result, release of pro-inflammatory cytokines is virtually absent. Thus, HPV efficiently evades innate immune responses leaving the adaptive immune system largely ignorant of the infection.[15]

Replication of the HPV genomes requires S-phase competent cells, providing enzymes necessary for DNA synthesis. Activation of the cell cycle in differentiated keratinocytes is accomplished by the viral E6 and E7 proteins that inhibit tumor suppressors p53 and members of the pRb (retinoblastoma- susceptibility protein) family, respectively.[16] Transcription of these oncogenes is controlled by upstream long control region (LCR), containing the E6/E7 promoters and a number of binding sites for both viral and cellular transcription factors.[17] Inhibition of pRb by E7 results in activation of transcription factor E2F, including expression of genes for DNA replication. In response to unscheduled DNA synthesis in differentiated cells, p53 is usually activated to induce cell cycle arrest or apoptosis. E6 mediated inhibition of p53 prevents apoptosis and permits continuous DNA replication and cell division.[17] These mechanisms are potentially oncogenic, since affected cells are no longer able to initiate growth arrest or apoptosis in response to DNA damage. Notably, only HR-HPV E6 and E7 can bind to and induce degradation of p53 and pRB.[18] In contrast, LR-HPV E6 and E7 only bind pRB and p53, but fail to induce degradation.[19] Although binding is weaker with HR-HPV oncogenes, interactions of LR-HPV E6 and E7 probably also abrogate pRB and p53 functions. Such interactions of LR-HPV with p53 and pRb members are likely to be important for dysregulation of the cell cycle and survival of infected cells. In addition, LR-HPV may prevent apoptosis independent of p53 via E6- mediated degradation of pro-apoptotic protein *Bak*.[20]

8.5.2 Chemistry

The primary drug substance in Veregen is sinecatechins, which is a partially purified fraction of the water extract of green tea leaves from *Camellia sinensis (L.) O Kuntze*, and is a mixture of catechins (Figure 8.1) and other green tea components. Catechins constitute 85%–95% (by weight) of the total drug substance which includes more than 55% epigallocatechin gallate (EGCg),

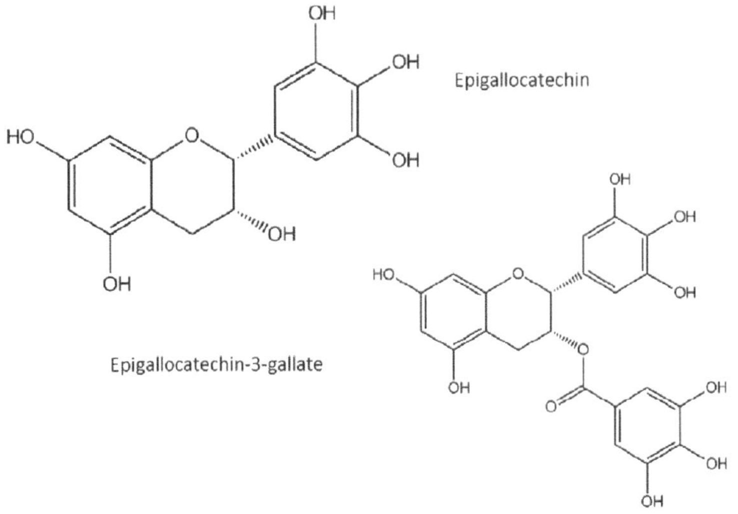

Figure 8.1 *Catechins are major bioactive principles in Veregen (Sinecatechins) botanical drug approved for the topical treatment of external genital and perianal warts.*

other catechin derivatives such as epicatechin (EC), Epigallocatechin (EGC), and epicatechin gallate (ECg), and some additional minor catechin derivatives, such as gallocatechin gallate (GCg), gallocatechin (GC) catechin gallate (Cg), and catechin (C). In addition to the known catechin components, it also contains gallic acid, caffeine, and theobromine, which constitute about 2.5% of the drug substance. The remaining amount of the drug substance contains unidentified botanical constituents derived from green tea leaves.

8.5.3 Preclinical characterization

Veregen exhibited no mutagenic markers, being negative in the Ames test, chromosome aberration assay, rat micronucleus assay and UDS test, but positive in the mouse lymphoma assay when used in doses up to 1000 mg/kg/day.[21] When orally administered at doses up to 500 mg/kg/day, Veregen was not associated with either neoplastic or non-neoplastic lesions in the tissues examined. Veregen was not teratogenic in rats and rabbits.[21] Daily vaginal administration of Veregen to rats during breeding and gestation did not cause adverse effects on mating performance and fertility at an approximate dose of 150 mg/rat/day. Furthermore, there were no adverse effects on embryo-fetal development in rats and rabbits, regardless of administration route.[21]

8.5.4 Pharmacokinetics

Systemic exposure to EGCg, EGC, ECg, and EC were evaluated following either topical application of Veregen to subjects with external genital and perianal warts (250 mg applied 3 times a day for seven days) or following oral ingesting of green tea beverage (500 ml ingested 3 times a day for seven days). Following topical application of Veregen, plasma concentrations in all 4 catechins were below the limit of quantification (<5 ng/ml) on day 1. After application of Veregen for seven days, plasma EGC, ECg, and EC concentrations were below the limit of quantification, while the plasma concentration of EGCg was measureable in 2 out of 20 subjects. The mean maximal plasma concentration of EGCg was 10.1 ng/ml. Oral ingestion of green tea beverage resulted in measurable concentration of EGCg in all subjects both on day 1 and day 7, with mean plasma concentration being 23.0 ng/ml.[7]

8.5.5 Clinical pharmacology

Three randomized, double-blind, placebo-controlled trials were conducted to determine the efficacy and safety of the catechin, Veregen, given as a 10% and 15% ointment. These studies sought to demonstrate the efficacy of each of the two Veregen ointment formulations over placebo with respect to the complete clearance rates of all baseline and new warts.[22] The primary efficacy outcome measure was the response rate defined as the proportion of subjects with complete clinical (visual) clearance of all external genital and perianal warts, baseline and new, by week 16.[7] The most affected areas in all studies were the penile shaft and the vulva.[20,22–24]

The first multicenter Phase II/Phase III combined study was reported by in 2007 and enrolled patients in 28 hospitals and practices in Germany and Russia. Eligible patients were required to be above the age of 18 with between 2 and 30 external angiogenital warts and a total wart area between 12 and 600 mm². Patients were asked to withhold any other treatment of angiogenital warts or Acyclovir/immunosuppressives 30 days prior to enrollment and have no record of HIV infection.[23] A total of 125 men and 117 women met the inclusion criteria and were randomly assigned to one of three treatment groups. 80 patients (42M/38F) were assigned to receive Veregen (Polyphenon E) 15% ointment, 79 patients (41M/38) assigned to receive Veregen 10% ointment, and 83 patients (42M/41F) were assigned to the placebo group. Treatment was administered topically, 3 times a day for up to 12 weeks or until complete clearance of baseline warts, with a 12-week treatment free follow-up period for complete responders. 90% of patients in all three groups had complete clearance during the 12 week treatment period. Mean time to complete clearance of all baseline warts in all groups was 10.6 ± 2.6 weeks.[23]

The next randomized, double-blind, 3 arm parallel group, vehicle controlled multicenter Phase III trial enrolled 226 women and 277 men from 46 dermatologic, gynecologic, and urologic centers throughout Europe and South Africa. The placebo-controlled trial studied the 15% and 10% Veregen ointments, and the results were reported in 2007.[24] The median wart area was 51 mm² and the median wart number was 6. A clearance of greater than 50% was achieved in 77.3% 78.0% and 52.9% of all subjects in the Veregen 15%, 10%, and placebo arms, respectively. Women responded better than men, with 45% of men in both active treatment groups achieving complete clearance of all warts. The median time to complete clearance was estimated at 16.1 weeks for both the 15% and 10% ointments, and 16.7 weeks for placebo (P < .001). Adverse events other than mild local reactions that were probably related to study medication were reported by four patients, including moderate balanitis, severe herpes simplex, mild lymphadenitis, and severe phimosis, all in the Veregen 15% ointment group.[24]

The third trial randomly assigned 502 subjects (258M/244F) from 50 health centers in the United States, Latin America, and Romania to the same three treatment arms reported by earlier.[25] One hundred and eleven subjects (57.2%) in the 15% ointment arm, 111 subjects (56.3%) in the 10% ointment arm, and 35 subjects (33.7%) in the placebo arm achieved complete clearance of all external genital warts (P < .001).[25] Compared with male patients, the proportion of female patients with complete clearance of all warts was higher in all three arms during the 12-week follow-up period. Recurrence occurred in 6.5%–8.8% of patients in all three treatment groups during the 12-week follow up period. Severe adverse events, comprising lymphadenitis, skin ulcer, vulvitis, and vulvovaginitis, were reported for five patients treated with 15% ointment, and for two treated with 10% ointment.[22,25]

8.5.6 Totality of evidence

As a naturally occurring mixture in which the active components are not well defined, the identified major and minor chemical components in Veregen need to be monitored and controlled for each marketed product batch. In the absence of data correlating chemical properties and clinical response, the acceptance ranges for these components were primarily established based on their levels observed in the multiple batches tested in clinical studies.[26] Significant variations of catechins and other chemical components have been identified from the tea leaves of different cultivars, and botanical products can have significant bath-to-batch variability (e.g., in total catechins and in ratios of different catechins).[26] In addition, although the majority of components can be adequately characterized and quantified, there may still be residual uncertainties about the chemical nature of minor components in Veregen. Therefore, to ensure consistent quality for Veregen, FDA considered two important pieces of information from the application, the variability in product batches due to collection site differences and a multiple dosing scheme.[26]

Therefore, the clinical studies performed with Veregen demonstrate good therapeutic efficacy and safety in therapy of genital and perianal warts with complete clearance rates of >50%. Sinecatechins not only reduced baseline warts but are also active against newly developed warts during treatment. As a possible advantage over Imiquimod and Podophyllotoxin, the recurrence rates after Veregen treatment was lower. Although not yet demonstrated, regression of wart lesions is likely to be caused by activation of cellular immune reactions, induction of cell cycle arrest and apoptosis, as well as inhibition of HPV transcription. This combination of molecular activities may expedite elimination of virus-infected cells of both clinical and sub-clinical lesions.[12]

8.6 Crofelemer (Fulyzaq): oral formulation of purified proanthocyanidin oligomers from the bark latex of the amazonian tree *Croton lechleri*

Fulyzaq, approved in 2012, is the first orally administered botanical drug approved by the FDA. It is also the first, and so far only, anti-diarrheal for the symptomatic relief of non-infectious diarrhea in adult patients with HIV/AIDS on anti-retroviral therapy. The active ingredients are oligomeric proanthocyanidins extracted from the bark latex of the *Croton lechleri* tree (Table 8.2), prescribed as a delayed-release tablet at a dose of 125 mg taken twice a day. Fulyzaq acts as an anti-secretory agent via its minimal systemic absorption following oral administration, which allows it to act locally in the gastrointestinal tract to block two principal chloride ion channels in the luminal membrane of the enterocyte.[27]

The exact mechanism of action for Fulyzaq is unknown, but a number of studies have proposed plausible mechanisms by which Fulyzaq could reduce secretions from the intestinal membrane. One study demonstrated that Fulyzaq inhibited the cyclic adenosine monophosphate (cAMP)-stimulated CFTR chloride channel located on the intestinal apical membrane, as well as the CaCC located on the intestinal epithelial membrane.[28] Both of these chloride channels regulate chloride and fluid secretion in the intestine and activation of both increases chloride and fluid secretion from the gastrointestinal tract, contributing to secretory diarrhea. Since Fulyzaq inhibits both of these channels, chloride secretion is decreased, causing both stool weight and frequency to be reduced and leading to relief of diarrhea. These results were corroborated in other independent studies investigating the use of Fulyzaq against different types of secretory diarrhea.[29-32]

Fulyzaq has also been highly active against diarrhea caused by particular bacterial species because of its inhibition of the CFTR chloride channel. *Escherichia coli* and *Vibrio cholera* produce enterotoxins that cause an increase in cAMP production. The elevated levels of cAMP simulate the CFTR chloride channel and increase chloride and fluid secretion, but these are blocked by Fulyzaq's proanthocyanidin oligomers.[30] Fulyzaq also shows antiviral activity against laboratory identified bacterial strains, including respiratory syncytial virus, influenza A virus, parainfluenza virus, herpes virus 1 and 2 and hepatitis A and B. This activity appears to develop from Fulyzaq's ability to bind to the viral envelope, preventing viral attachment and penetration of the host cell.[33,34]

8.6.1 Standard indication: HIV-associated diarrhea

Secretory diarrhea is commonly comorbid with human immunodeficiency virus (HIV). Diarrhea can affect patients at any stage of illness, with up to 60% of patients with HIV reporting symptoms of diarrhea.[35,36] The type of diarrhea that develops in HIV patients generally has secretory properties. Secretory diarrhea results whenever there is an excess secretion of chloride ions followed by movement of sodium and water into the intestinal lumen. Increased secretion of chloride ions can occur when the cystic fibrosis transmembrane conductance regulator (CFTR) and calcium-activated chloride channels (CaCC) are overstimulated. HIV and various antiretroviral agents can activate these channels, leading to the development of secretory diarrhea.[28] *Cryptosporidium, Isospora belli, Microsporidia,* and *Mycobacterium avium-intracellulare,* as well as *Salmonella, Shigella,* and *Campylobacter* are all opportunistic pathogens capable of causing diarrhea[36,37] however, diarrhea caused by opportunistic pathogens was more common before antiretroviral therapy.

The goal of antiretroviral therapy is to completely suppress viral replication via the introduction of highly active antiretroviral therapy (HAART), and HIV patients are experiencing better clinical outcomes with improved survival. Nonetheless, HIV-associated diarrhea remains a common side effect of the HAART, along with HIV enteropathy, autonomic neuropathy, chronic pancreatitis, and exocrine insufficiency.[29,38] HIV enteropathy is a form of diarrhea that can occur during any stage of HIV infection and can last for months. It often arises without a clear infectious cause and it comprises a variety of gastrointestinal illnesses including diarrhea, GI inflammation, increased intestinal permeability, and malabsorption. Especially during the acute phase of HIV infection, gut lymphoid tissue becomes one of the major sites for HIV replication, which can lead to a significant loss of CD4+ T-cells. When the GI tract is infiltrated by lymphocytes, structural changes due to inflammatory processes can occur, like villous atrophy and crypt hyperplasia.

HIV causes a variety of other changes that could be linked to secretory diarrhea. Local activation of immune cells by HIV leads to the release of pro-inflammatory signals like interleukin-1, tumor necrosis factor α, and macrophage inflammatory proteins 1α and 1β in the GI tract.[38] The HIV transactivating factor protein can stimulate calcium-activated chloride ion secretion in enterocytes and colonic mucosa. This protein may also inhibit the proliferation of enterocytes, leading to induction of apoptosis in these cells. Glycoprotein 120, an HIV envelope protein, causes increased calcium concentration in enterocytes that leads to tubulin depolymerization and inadequate epithelial ion balance.[39] The HIV protein R has also been shown to possess inflammatory properties. This protein also disrupts barrier function, which contributes to the development of HIV enteropathy. Through these mechanisms, HIV itself can lead to GI dysfunction and possible secretory diarrhea.

8.6.2 Chemistry

Fulyzaq is a natural compound isolated from the bark of the *Croton lechleri* tree from the *Euphorbiaceae* family. This tree is commonly found in the western Amazonian region of South America, where the red latex is commonly referred to as "dragons blood" or "sangre de drago."[40] Fulyzaq is an acid-labile, proanthocyanidin oligomer (Figure 8.2) with an average molecular weight of 2100 daltons. The monomeric components of the polyphenolic molecule include (+)-catechin, (–)-epichatechin, (+)gallocatechin and (–) galloepicatechin.[41]

8.6.3 Preclinical characterization

Fulyzaq exhibited no mutagenic markers, being negative in the Ames test, chromosome aberration assay at a dose of 600 ug/ml and rat bone marrow micronucleus study at a dose of 50 mg/kg.[42] Furthermore, at a dose of 738 mg/kg/day (177 times the recommended human daily dose) Fulyzaq had no effect on fertility or reproductive performance of male or female rats.[41,42] Additionally, Fulyzaq was not teratogenic at this dose.[42] Carcinogenicity studies were not conducted prior to NDA approval by the FDA, but appropriate studies will be completed post-marketing, as agreed in the pre-NDA meeting with the FDA.[42]

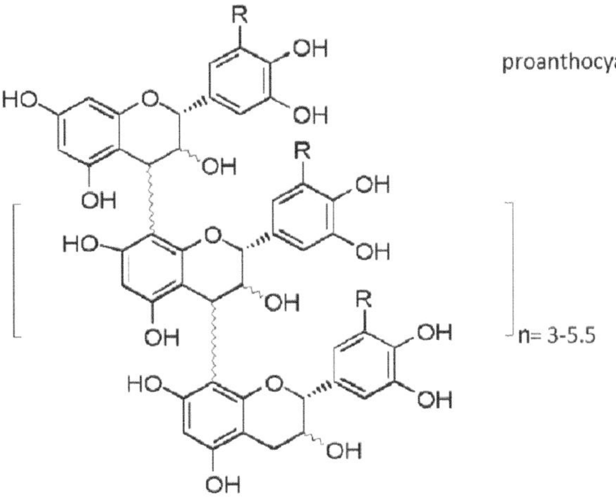

proanthocyanidin

n= 3-5.5

Figure 8.2 *Proanthocyanidins are major bioactive principles in Fulyzaq (Crofelemer) botanical drug approved for relieving symptoms of diarrhea in HIV/AIDS patients taking antiretroviral therapy.*

8.6.4 Pharmacokinetics

Clinical trials have used a range of doses between 125 and 500 mg every 6–12 hours. Recent FDA approval labeling for Fulyzaq recommends using the 125 mg delayed-release tablets twice a day for the treatment of symptomatic noninfectious diarrhea in adult HIV/AIDS patients on antiretroviral therapy.[41] Oral Fulyzaq has little to no systemic absorption. Plasma concentrations are undetectable after ingestion of Fulyzaq and metabolites have not been identified in the blood. Food does not affect the efficiency or absorption of Fulyzaq, as co-administration with a fatty meal did not result in increased systemic exposure. Although in vitro studies show that Fulyzaq has the potential to inhibit cytochrome P450 isoenzyme 3A and transporters MRP2 and OATP1A2 at concentrations expected in the gut, due to its minimal absorption Fulyzaq is unlikely to inhibit multiple cytochrome P450 isoenzymes.[41] Despite Fulyzaq's potential, there are no reported clinically relevant drug interactions. Specifically, no drug-drug interactions were found between Fulyzaq and antiretrovirals such as Nelfinavir, Zidovudine, and Lamivudine, although a 20% decrease in Lamivudine exposure was observed in patients receiving Fulyzaq 500 mg four times daily, but this was not considered to be clinically important.[41]

8.6.5 Clinical pharmacology

A Phase II, randomized, double-blind, placebo-controlled study was designed to evaluate the safety and efficacy of Fulyzaq for the treatment of HIV-associated diarrhea as assessed by stool weight and frequency. HIV patients between 18 and 60 years old and with chronic diarrhea were included. Chronic diarrhea was defined as having at least three soft or watery stools, with stool weight of more than 200 g/day. Eligible patients were also required to have a diagnosis of AIDS as defined by the CDC and be on appropriate antiretroviral therapy for at least 2 weeks before and throughout the duration of the clinical trial. Study patients discontinued all antidiarrheal agents 24 hours before the study began. Baseline stool weight and frequency were determined for each participant during a 24-hours observation period. A total of 51 patients were included in the study and were randomized to either the treatment or control group. Twenty-six people were assigned 500 mg (two 250 mg tablets) of Fulyzaq every six hours for four days, and 25 people were assigned two placebo capsules every six hours for four days.[32]

The mean baseline stool weight assessed during the 24-hours observation period was 914.8 g for the treatment arm and 813.9 g for the placebo arm. The mean baseline stool frequency during the 24-hours observation period was 5.2 stools for both groups.[32] Comparing day 4 results with baseline, patients treated with Fulyzaq had a significantly greater average reduction of stool weight and frequency compared with the placebo group. The mean reduction in stool weight from baseline to day 4 was 451.3 g/day for the treatment group compared to 150.7 g/day in the placebo group. The mean reduction in stool frequency from baseline to day 4 was three stools per day in the Fulyzaq group and two stools per day in the control group. Patients in the Fulyzaq group had a mean reduction in chloride concentration after four days, while the control group had a mean increase.[32] The results of this Phase II trial indicate that Fulyzaq could be useful in reducing stool weight and diarrhea. Another important observation from this study is that 77% of patients were on PI-based HAART. This further supports the role of Fulyzaq in the treatment of diarrhea in HIV patients whose diarrhea has multiple causes.

The ADVENT study was a Phase III, randomized double-blind, placebo-controlled multicenter trial designed to evaluate the efficacy of Fulyzaq in the treatment of secretory diarrhea in HIV-infected patients. For this study, secretory diarrhea was defined as persistently loose stools even with the regular use of antidiarrheal agents, or one or more watery stools per day without the use of antidiarrheal agents. Eligible patients were receiving stable antiretroviral therapy, had a history of diarrhea for at least one month, had CD4+ T cell counts greater than 100 cells/ul, and had no evidence of infection. Patients were excluded if they had a history of GI disease that caused diarrhea.[41,43]

The ADVENT study consisted of two stages. Stage 1 established optimal dosing, and stage 2 was used to assess safety and efficacy of Fulyzaq. Each stage included two phases, a 4-week placebo-controlled phase, and a five month treatment-extension phase where all the patients received Fulyzaq. A 10-day screen period during which all patients received placebo preceded the placebo-controlled phase. Randomization to the placebo-controlled phase occurred only if patients experienced one or more watery bowel movements per day on at least 5 of seven days of the screening period. The primary efficacy endpoint of the study was they proportion of patients who demonstrated a response to Fulyzaq; clinical response was defined as no more than two watery bowel movements per week for at least 2 of the four weeks of the placebo-controlled phase.[41,43]

A total of 374 patients were enrolled and randomized to each treatment arm: 236 in the Fulyzaq arm and 138 in the placebo arm. The median number of daily watery bowel movements was 2.5 per day. PI-based antiretroviral regimens were the most common in the treatment groups (64% in the Fulyzaq 125 mg twice daily group). Among these, 22% of patients in this group were receiving Lopinavir/ Ritonavir.[43] In stage 1, patients were randomized 1:1:1:1 to receive either Fulyzaq 125, 250, and 500 mg, or placebo twice daily. Results from stage 1 of the ADVENT study revealed an optimal dosing regimen of Fulyzaq 125 mg twice daily. Patients in this arm experienced a better clinical response than the control group. In stage 2, patients were randomized to receive either 125 mg of Fulyzaq twice daily or the placebo in the first phase. During this phase, 16.3% of patients in the Fulyzaq arm experienced a clinical response, compared with 11.4% in the placebo arm. During the second phase all patients received Fulyzaq. Combined data from both phases showed that 17.6% of patients in the Fulyzaq group demonstrated a significant clinical response versus only 8% in the placebo group. Patients who received placebo during the 4-week phase and then crossed over to Fulyzaq showed considerable improvement after one month of use. These patients were also found to have a greater chance of experiencing a clinical response in the remaining four months of the treatment period. Treatment response was consistent among pre-specified subgroups, including duration of diarrhea, baseline number of watery bowel movements, use of PIs, CD4+ count, and age. Fulyzaq was found to be less effective among African Americans when examining treatment effect consistency across race subgroups.[41,43] Results from the ADVENT study provide supportive evidence for FDA approval of Fulyzaq 125 mg delayed-release capsules twice daily for the symptomatic relief of noninfectious diarrhea in adults with HIV/AIDS treated with antiretroviral therapy.

8.6.6 Totality of evidence

Compared with sinecatechins (Veregen), the chemical characterization of Fulyzaq presented an even greater challenge. This drug consists of a mixture of oligomers that vary in composition, sequence, and length, which precluded adequate separations and quantification of proanthocyanadin oligomers based on multiple conventional methods of detection. Advanced chromatographic, spectroscopic, spectrometric, and acid hydrolysis methods were needed to provide a comprehensive characterization of Fulyzaq. These analytical methods collectively revealed extensive information on the structural signatures of Fulyzaq but they were ultimately considered insufficient to support the characterization and quality control of this complex botanical mixture.[44] Considering the degree of uncertainty regarding the chemical characterization of Fulyzaq, the FDA concluded that in addition to raw material control and clinical data from multiple doses and product batches, a clinically relevant bioassay to assess the drug product activity was needed before it could be approved.[26,28]

8.7 ERr 731 extract (Estrovera): oral formulation of hydroxystilbenes extracted from rhubarb roots

German clinicians have been recommending a purified, standardized extract of Siberian rhubarb (*Rheum rhaponticum L.*) known as Estrovera/ ERr 731 for long-term treatment of menopausal symptoms since 1993, and it was launched in the US market by Metagenics Inc in 2010. The active ingredients are hydroxystilbenes extracted from the root of Siberian rhubarb (*Rheum rhaponticium L.*)[45,46] (Table 8.2). Clinical studies have demonstrated that one tablet (4 mg) daily offers significantly effective relief versus placebo for the 11 most common menopausal symptoms, including hot flushes. The mechanism of action of ERr 731 is still not completely understood. Preliminary research suggests that Estrovera may act as a selective estrogen receptor modulator (SERM) for estrogen receptor β, which may explain its efficacy and safety profile.[47] Such activation may also alleviate symptoms of depression and anxiety by modulating neurotransmitter release associated with ER signaling.[48] ERr 731 constituents, rhapontigenin and desoxyrhapontigenin, have demonstrated inhibition of monoamine oxidase A with serotonin as a substrate. This effect has been suggested to favorably modulate serotonin and catecholamine metabolism to support a healthy mood and cognitive function.[49]

8.7.1 Standard indication: to relief menopause symptoms

Menopause is the clinical term used after menstruation has ceased for one year, after which women are considered postmenopausal. Perimenopause is used to describe the time leading up to the final menstruation, which is signaled by irregular menstrual bleeding, erratic hormone levels, and the onset of menopausal symptoms.[50] An estimated 6,000 American women transition to menopause every day, with 75% of women aged 50–55 years old assumed to be postmenopausal.[51] Alternatively, induced menopause is the cessation of menstruation caused by suppression of ovarian function either through surgical removal, pelvic radiation therapy, or chemotherapy. Women with induced menopause may suffer symptoms of greater intensity or frequency. Over 90% of women who undergo surgical removal suffer from hot flushes and other symptoms that can be chronic and severe.[52]

The most common symptoms typically fall into 11 categories, with hot flushes being the most common and potentially debilitating. Nearly 80% of women in Western countries suffer from hot flushes, with 30% reporting severe and frequent enough hot flushes to seriously affect quality of life.[53] They can significantly affect daily functioning and sleep, as well as reported state of health. Sleep disturbances are the fourth most frequent menopausal complaint, with up to 60% of women reporting trouble sleeping. Additionally, more than 40% of women experience physical and mental exhaustion and cite forgetfulness as a menopausal symptom.[50,51] Declines in cognitive function have been linked to sleep disturbances during menopause.[54]

While the physiology of hot flush vasomotor response is unknown, it appears to be a result of dysfunction of the thermoregulatory centers influenced by the hypothalamus.[50,54-56] Furthermore, estrogen withdrawal, rather than low circulating estrogen, is postulated to be the primary cause of hot flushes. This is supported by women who suddenly discontinue HT.[50,54-56] Norepinephrine is suggested to be the primary neurotransmitter influencing these thermoregulatory changes.[55,57] However, hormonal fluctuations in progesterone, as well as estrogen, are another purported systemic influence for hot flushes.[56] A decline in progesterone, which exerts a sedative and anti-excitatory activity by modulating GABA receptors, may also contribute to anxiety and altered sleep patterns.[58] Furthermore, when ovarian follicles fail to secrete estrogen to provide negative feedback for regular cycling, pituitary gonadotropin increases and leads to increased levels of luteinizing hormone (LH).[47,57] Though no causal relationship between the level of circulating LH and hot flushes has been demonstrated, hot flush occurrence has been correlated with pulses in LH levels. These pulses may also involve the thermoregulatory response via their effect on hypothalamic neurons.[50]

8.7.2 Chemistry

ERr 731 is a unique phytoestrogen mixture extracted from the root of Siberian rhubarb (*Rheum rhaponticium L.*).[45,46] The main active constituents of ERr 731 is the glycoside rhaponticin, followed by desoxyrhaponiticin. The metabolites, rhapontigenin and desoxyrhapontigenin, comprise about 5% of the extract (Figure 8.3). In plants, these compounds are synthesized to protect against viral and microbial attack, disease, and ultraviolet exposure. These hydroxystilbene compounds are structurally related to resveratrol, which has also demonstrated SERM activity.[59,60] Siberian rhubarb is different than other medicinal rhubarbs from other parts of the world, with fewer anthraquinones (known for their laxative effect) and a higher concentration of hydroxystilbene compounds near the root.[45,46]

8.7.3 Preclinical characterization

Long-term studies in rats have shown no negative effects in bone density or uterine tract changes.[61] Dogs treated with the no-adverse-effect level (NOAEL) of 1000 mg/kg/day for 13 weeks showed no pathological changes to the organs of the uterine tract or changes in uterine weight.[62]

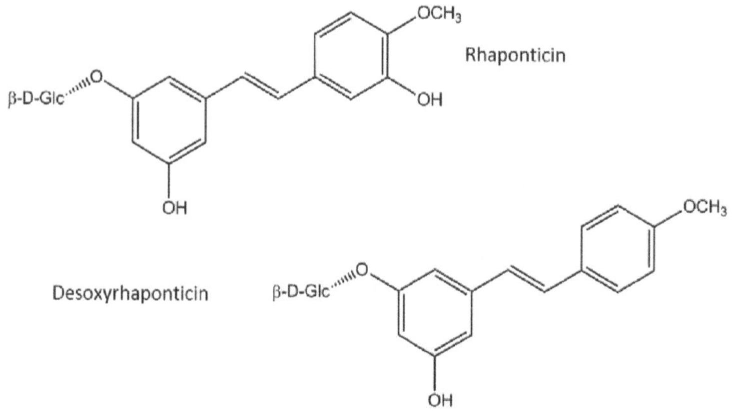

Figure 8.3 *Hydroxystilbenes are the major chemical constituents and putative bio-active principles in Estrovera nutraceutical product marketed to relieve common symptoms of hot flashes and night sweat associated with the menopause.*

8.7.4 Pharmacokinetics

A dose of 20 mg/kg of *R. rhaponticum* hydroxystilbenes was administered to mice via an intraperitoneal injection and blood collections were done 5,10,30, 45 minutes and 1,2,3,4, and 24 hours after administration.[63] In the blood, seven initial trans-hydroxystilbenes and five of their metabolites were identified. Most of the trans-hydroxystilbenes found had peaked in the bloodstream 10 minutes after administration and metabolites peaked 15 minutes after administration.[63] There was no measurable concentration of trans-hydroxystilbenes three hours after administration. No bioaccumulation of parent polyphenols or their metabolites was observed in the course of chronic administration.[63]

8.7.5 Clinical pharmacology

Although not approved by the FDA, clinical studies were designed to investigate and minimize factors such as major vasomotor triggers and pre-existing conditions that may influence outcomes. The primary outcome criterion for efficacy used was the change of the Menopause Rating Scale (MRS) between treatment and placebo groups.[45] The exclusion criteria for the following studies are as follows: regular cycles in the last three months, Pap smear of class III/IV hyperplasia, and BMI below 18 or above 30, or abnormal eating habits.[45–47,53] Subjects were also excluded if they had a history of Type 2 diabetes or prescription corticosteroids, as well as previous or existing thromboembolic disease, insufficiently controlled hypertension, or hypertensive medication. Individuals with a history of smoking, drug or alcohol abuse, previous or existing psychiatric disorders, or high intake of caffeine (>500 mg/day) were also excluded.[53]

In the first study, 109 symptomatic perimenopausal women received either ERr 731 (n = 54) or placebo (n = 55) for 12 weeks. The ERr 731 group showed significant improvements in 11 common menopausal symptoms. At four weeks there was a significant decrease in the number and severity of hot flushes compared to the placebo group (P < 0.0001), along with significant decrease in Hamilton Anxiety Scale (HAMA) total score for somatic and cerebral anxiety, and a general improvement in total Women's Health Questionnaire Score (WHQ), including measurements in anxiety and poor mood.[53] At 12 weeks, the ERr 731 group demonstrated a significant decrease in total Menopause Rating Scale II (MRS II) score, as well as significant decreases in all 11 individual symptom scores compared to placebo.[53] From week 4, there were improvements in HAMA scores as well as the MHQ score.[53]

In another 12 week study, 112 symptomatic perimenopausal women between the age of 45 and 55 were given one tablet per day of Estrovera 4 mg (n = 56) or placebo (n = 56). At 12 weeks, subjects given Estrovera showed a significant reduction in the number of daily hot flushes, from a median of 12–2.[45] Based on the Hot-Flush-Weekly-Weighted Score (HFWWS), this decrease in hot flushes is comparable to those reported for an ultra-low dose of HT. Additionally, those with the most severe hot flushes received the most relief from Estrovera intervention. There was also a significant reduction of the MRS total score, from an average of 27–12.4.[45] Estrovera subjects also showed significant reductions in each of the 11 individual MRS scores.[45]

In another multi-center clinical trial, 252 women between the ages of 39 and 71 were recruited from 70 gynecological practices in an open observational study to receive ERr 731 at various doses for six months.[46] During the first three months, 243 participants took one tablet (4 mg) daily, 13 took two tablets daily, and 1 woman took four tablets daily. Over the course of the entire six months study, 228 women took one tablet daily, and 6 took two tablets daily, as recommended. For the majority of women, one tablet was enough to significantly relieve menopausal symptoms.[46] Subjects showed a significant decrease in MRSII total score from an average 14.5–6.5 points and reported a notable improvement in quality of life most noticeable in women who started with a MRSII score above 18.[46]

Finally, in a 108-week study, 80 subjects from the first 12 weeks study were followed for observational studies along with a placebo group. In the first 48 weeks, subjects received either ERr 731 (n = 39) or placebo (n = 41). In the second 48 weeks, 41 women were given ERr 731 (23 from the last ERr 731 group and 28 from the last placebo group). ERr 731 demonstrated a further decrease or sustained alleviation of menopausal symptoms at 60 and 108 weeks.[47] At 108 weeks, all subjects (now receiving ERr 731, but with varying lengths of treatment) had an average of less than 1.4 slight hot flushes per day.[47]

8.7.6 Totality of evidence

All recommended therapies for menopause (hormonal or non-hormonal) are centered on positive lifestyle changes that can not only ease the degree of symptomology, but also influence the rate of transition. ERr 731 is perhaps the most thoroughly tested phytoestrogen SERM to date that offers a more natural approach to relieving menopausal symptoms, including hot flushes. Published toxicology and clinical studies suggest reliable efficacy and predicted long-term safety with no associated serious adverse events reported to date. When ERr 731 is used in conjunction with a patient-centered approach to menopausal relief, it may offer positive clinical outcomes for women in various stages of menopausal transition.[47,50,52]

Despite the clinical data already collected for ERr 731, it has not been submitted as an IND or an NDA, and can make no claims to treat a disease. Instead, Metagenics, the company responsible for the production of ERr 731, has created a dietary supplement and must rely directly on the menopausal population and health practitioners to advertise its effectiveness. Thus, foregoing FDA endorsements and allowing personal recommendation and popular knowledge to drive sales while maintaining strict standards of evidence-based preclinical and clinical programs to evaluate botanical product safety and efficacy, seems like a less-costly and therefore more attractive strategy to bringing botanical products to the market.

8.8 Conclusions

Drawing conclusions on the prospects of botanical drugs in general is still premature for a number of reasons, including variable complex mixtures, multiple plant combinations, extensive previous human use, and availability as dietary supplements before approval as drugs. In fact, of the 282 pre-INDs and INDs submitted between 1999 and 2007, only 36% were multi-plant combinations, reflecting the difficulties in working with more complex preparations.[1] Moreover, as of 2012, the strong interest in developing botanical drugs resulted in more than 500 pending FDA applications, but progress in developing new drugs from botanicals has been slow, with only two botanical NDAs to date. While extensive history of prior human use may reduce the early requirements for nonclinical pharmacology and toxicology for the initial (Phase I and II) clinical studies during the development of the botanical drug, these data will be required for the subsequent Phase III and the final marketing approval.

In this situation, the opportunities for sponsors with limited R&D capabilities and reduced promotional budgets are few: more effective resources management (flexibility, opportunistic approach, focus on niches) but also finding less conventional regulatory approaches and exploring alternative pathways to market. In fact, to bypass FDA regulation of the safety of functional substances, it is thus in a manufacturer's interest to try to market a botanical product as a dietary supplement rather than as a conventional food with added ingredients or a botanical drug. By marketing botanical products as dietary supplements, sponsors can avoid having to prove that added ingredients are GRASE and when questions arise, such products can stay on the market until the FDA proves in court that they may be harmful. Implementation of the strict evidence-based preclinical and clinical programs to evaluate safety and efficacy of the botanical products in either category, however, seems to highlight a prominent unifying trend that will drive the development of future botanical products.

References

1. Chen, S. T. et al. New therapies from old medicines. *Nat. Biotech.* 26, 1077–1083 (2008).
2. Center for Drug Evaluation and Research. *Botanical Drug Development: Guidelines for Industry.* Silver Spring, MD, (2015).
3. The Drug Approval Process. Available at: http://www.medscape.com/viewarticle/405869_4. (Accessed: March 1st, 2016).
4. The Drug Development Process > Step 3: Clinical Research. Available at: http://www.fda.gov/ForPatients/Approvals/Drugs/ucm405622.htm. (Accessed: March 11, 2016).
5. Mishra, B. B. & Tiwari, V. K. Natural products: An evolving role in future drug discovery. *Eur. J. Med. Chem.* 46, 4769–4807 (2011).
6. Shimizu, M. & Weinstein, I. B. Modulation of signal transduction by tea catechins and related phytochemicals. *Mutat. Res. Mol. Mech. Mutagen.* 591, 147–160 (2005).
7. Treatment for Genital and Perianal Warts—VEREGEN® Ointment. Available at: http://veregen.com/. (Accessed: January 13, 2016).
8. Hoy, S. M. Polyphenon E 10% Ointment: In Immunocompetent Adults with External Genital and Perianal Warts. *Am. J. Clin. Dermatol.* 13, 275–281 (2012).
9. Burchell, A. N., Winer, R. L., de Sanjosé, S. & Franco, E. L. Chapter 6: Epidemiology and transmission dynamics of genital HPV infection. *Vaccine* 24, S52–S61 (2006).
10. Steben, M. & Duarte-Franco, E. Human papillomavirus infection: Epidemiology and pathophysiology. *Gynecol. Oncol.* 107, S2–S5 (2007).
11. Kaufman, R. H., Adam, E. & Vonka, V. Human papillomavirus infection and cervical carcinoma. *Clin. Obstet. Gynecol.* 43, 363 (2000).
12. Stockfleth, E. & Meyer, T. The use of sinecatechins (polyphenon E) ointment for treatment of external genital warts. *Expert Opin. Biol. Ther.* 12, 783 (2012).
13. Wiley, D. et al. External genital warts: Diagnosis, treatment, and prevention. *Clin. Infect. Dis.* 35, S210–S224 (2002).
14. Doorbar, J. The papillomavirus life cycle. *J. Clin. Virol.* 32, 7–15 (2005).
15. Stanley, M. Immune responses to human papillomavirus. *Vaccine* 24, S16–S22 (2006).
16. Zur Hausen, H. Papillomaviruses and cancer: From basic studies to clinical application. *Nat. Rev. Cancer* 2, 342–350 (2002).
17. Hoppe-Seyler, F. & Butz, K. Cellular control of human papillomavirus oncogene transcription. *Mol. Carcinog.* 10, 134–141 (1994).
18. Dyson, N., Howley, P. M., Munger, K. & Harlow, E. The human papilloma virus-16 E7 oncoprotein is able to bind to the retinoblastoma gene product. *Science* 243, 934–937 (1989).
19. Boyer, S. N., Wazer, D. E. & Band, V. E7 protein of human papilloma virus-16 induces degradation of retinoblastoma protein through the ubiquitin-proteasome pathway. *Cancer Res.* 56, 4620–4624 (1996).
20. Thomas, M. & Banks, L. Human papillomavirus (HPV) E6 interactions with bak are conserved amongst E6 proteins from high and low risk HPV types. *J. Gen. Virol.* 80, 1513–1517 (1999).
21. Veregen Pharmacology Review. (2006).

22. Meltzer, S. Green tea catechins for treatment of external genital warts. *Am. J. Obstet. Gynecol.* 200, 233.e1–233.e7 (2009).
23. Gross, G. et al. A randomized, double-blind, four-arm parallel-group, placebo-controlled Phase II/III study to investigate the clinical efficacy of two galenic formulations of Polyphenon® E in the treatment of external genital warts. *J. Eur. Acad. Dermatol. Venereol.* 21, 1404–1412 (2007).
24. Stockfleth, E. et al. Topical Polyphenon E in the treatment of external genital and perianal warts: A randomized controlled trial. *Br. J. Dermatol.* 158, 1329–1338 (2008).
25. Tatti, S. et al. Polyphenon E: A new treatment for external anogenital warts. *Br. J. Dermatol.* 162, 176 (2010).
26. American Association for the Advancement of Science. The art and science of traditional medicine part 2: Multidisciplinary approaches for studying traditional medicine. *Science* 347, 337–337 (2015).
27. Frampton, J. E. Crofelemer: A review of its use in the management of non-Infectious diarrhoea in adult patients with HIV/AIDS on antiretroviral therapy. *Drugs* 73, 1121–1129 (2013).
28. Tradtrantip, L., Namkung, W. & Verkman, A. S. Crofelemer, an antisecretory antidiarrheal proanthocyanidin oligomer extracted from croton lechleri, targets two distinct intestinal chloride channels. *Mol. Pharmacol.* 77, 69–78 (2010).
29. MacArthur, R. D. & DuPont, H. L. Etiology and pharmacologic management of noninfectious diarrhea in HIV-infected individuals in the highly active antiretroviral therapy era. *Clin. Infect. Dis. Off. Publ. Infect. Dis. Soc. Am.* 55, 860 (2012).
30. Mangel, A. W. & Chaturvedi, P. Evaluation of crofelemer in the treatment of diarrhea-Predominant irritable bowel syndrome patients. *Digestion* 78, 180–186 (2009).
31. S. E. Gabriel et al. A novel plant-derived inhibitor of cAMP-mediated fluid and chloride secretion. *Am. J. Physiol.—Gastrointest. Liver Physiol.* 276, 58–63 (1999).
32. Holodniy, M. et al. A double blind, randomized, placebo-controlled phase II study to assess the safety and efficacy of orally administered SP-303 for the symptomatic treatment of diarrhea in patients with AIDS. *Am. J. Gastroenterol.* 94, 3267–3273 (1999).
33. Gilbert, B. E., Wyde, P. R., Wilson, S. Z. & Meyerson, L. R. SP-303 Small-particle aerosol treatment of influenza a virus infection in mice and respiratory syncytial virus infection in cotton rats. *Antiviral Res.* 21, 37–45 (1993).
34. Wyde, P. R., Ambrose, M. W., Meyerson, L. R. & Gilbert, B. E. The antiviral activity of SP-303, a natural polyphenolic polymer, against respiratory syncytial and parainfluenza type 3 viruses in cotton rats. *Antiviral Res.* 20, 145–154 (1993).
35. DuPont, H. L. Acute infectious diarrhea in immunocompetent adults. *N. Engl. J. Med.* 370, 1532 (2014).
36. Patel, T. S., Crutchley, R. D., Tucker, A. M., Cottreau, J. & Garey, K. W. Crofelemer for the treatment of chronic diarrhea in patients living with HIV/AIDS. *HIVAIDS Auckl. NZ* 5, 153 (2013).
37. Kartalija, M. & Sande, M. A. Diarrhea and AIDS in the era of highly active antiretroviral therapy. *Clin. Infect. Dis.* 28, 701–705 (1999).
38. Macarthur, R. D. Management of noninfectious diarrhea associated with HIV and highly active antiretroviral therapy. *Am. J. Manag. Care* 19, s238 (2013).
39. Brenchley, J. M. & Douek, D. C. HIV infection and the gastrointestinal immune system. *Mucosal Immunol.* 1, 23–30 (2008).
40. Fischer, H. et al. A novel extract SB-300 from the stem bark latex of Croton lechleri inhibits CFTR-mediated chloride secretion in human colonic epithelial cells. *J. Ethnopharmacol.* 93, 351–357 (2004).
41. Fulyzaq (crofelemer) [precribing information]. (2013). https://shared.salix.com/shared/pi/fulyzaq-pi.pdf.
42. Fulyzaq Pharmacology Review. (2012). https://www.clinicaltrials.gov/ct2/show/NCT00547898?term=crofelemer.

43. MacArthur, R. D. et al. Efficacy and safety of crofelemer for noninfectious diarrhea in HIV-seropositive individuals (ADVENT trial): A randomized, double-blind, placebo-controlled, two-stage study. *HIV Clin. Trials* 14, 261–273 (2013).

44. Kennedy, J. A. & Jones, G. P. Analysis of proanthocyanidin cleavage products following acid-catalysis in the presence of excess phloroglucinol. *J. Agric. Food Chem.* 49, 1740–1746 (2001).

45. Kaszkin-Bettag, M., Ventskovsky, B. & Solskyy, S. Confirmation of the efficacy of ERr 731 in perimenopausal women with menopausal symptoms. *Altern. Ther. Health Med.* 15, 24–34 (2009).

46. Kaszkin-Bettag, M., Beck, S., Richardson, A., Heger, P. W. & Beer, A.-M. Efficacy of the special extract ERr 731 from rhapontic rhubarb for menopausal complaints: A 6-month open observational study. *Altern. Ther. Health Med.* 14, 32–38 (2008).

47. Hasper, I. et al. Long-term efficacy and safety of the special extract ERr 731 of Rheum rhaponticum in perimenopausal women with menopausal symptoms. *Menopause* 16, 117–131 (2009).

48. Wober, J. et al. Activation of estrogen receptor-β by a special extract of Rheum rhaponticum (ERr 731®), its aglycones and structurally related compounds. *J. Steroid Biochem. Mol. Biol.* 107, 191–201 (2007).

49. Kaszkin-Bettag, M. et al. The special extract ERr 731 of the roots of Rheum rhaponticum decreases anxiety and improves health state and general well-being in perimenopausal women. *Menopause* 14, 270–283 (2007).

50. Umland, E. M. Treatment strategies for reducing the burden of menopause-associated vasomotor symptoms. *J. Manag. Care Pharm.* 14, 14–19 (2008).

51. Nelson, H. D. et al. Management of menopause-related symptoms: Summary. *Evid. Rep. Technol. Assess.* 120, 1–6 (2005).

52. Hickey, M. et al. Practical clinical guidelines for assessing and managing menopausal symptoms after breast cancer. *Ann. Oncol.* 19, 1669–1680 (2008).

53. Heger, M. et al. Efficacy and safety of a special extract of Rheum rhaponticum (ERr 731) in perimenopausal women with climacteric complaints: A 12-week randomized, double-blind, placebo-controlled trial. *Menopause* 13, 744–759 (2006).

54. Maki, P. M. et al. Objective hot flashes are negatively related to verbal memory performance in midlife women. *Menopause N. Y.* 15, 848 (2008).

55. Nelson, H. D. et al. Nonhormonal therapies for menopausal hot flashes: Systematic review and meta-analysis. *J. Am. Med. Assoc.* 295, 2057–2071 (2006).

56. Albertazzi, P. Alternatives to estrogen to manage hot flushes. *Gynecol. Endocrinol.* 20, 13–21 (2005).

57. Shanafelt, T. D., Barton, D. L., Adjei, A. A. & Loprinzi, C. L. Pathophysiology and treatment of hot flashes. *Mayo. Clinic. Proc.* 77, 1207–1218 (2002).

58. Sherwin, B. Progestogens used in menopause. Side effects, mood and quality of life. *J. Reprod. Med.* 44, 227–232 (1999).

59. Geller, S. E. & Studee, L. Botanical and dietary supplements for mood and anxiety in menopausal women. *Menopause* 14, 541–549 (2007).

60. Geller, S. E. et al. Safety and efficacy of black cohosh and red clover for the management of vasomotor symptoms: A randomized controlled trial. *Menopause N. Y.* 16, 1156 (2009).

61. Keiler, A. M., Papke, A., Kretzschmar, G., Zierau, O. & Vollmer, G. Long-term effects of the rhapontic rhubarb extract ERr 731® on estrogen-regulated targets in the uterus and on the bone in ovariectomized rats. *J. Steroid Biochem. Mol. Biol.* 128, 62–68 (2012).

62. Kaszkin-Bettag, M., Richardson, A., Rettenberger, R. & Heger, P. Long-term toxicity studies in dogs support the safety of the special extract ERr 731 from the roots of Rheum rhaponticum. *Food Chem. Toxicol.* 46, 1608–1618 (2008).

63. Raal, A. et al. Trans-resveratrol alone and hydroxystilbenes of rhubarb (Rheum rhaponticum L.) root reduce liver damage induced by chronic ethanol administration: A comparative study in mice. *Phytother. Res. PTR* 23, 525–532 (2009).

9

Botanical Drug Products and Rare Diseases

Jayant N. Lokhande and Sonali Lokhande

Contents

R are or orphan diseases, as the name indicates, are a group of disorders that affect a small section of the population. The definition of orphan disease varies in each country and depends upon the regulatory guidelines and healthcare policy in that specific region of the world. The rare disease patient populations are defined in law as [1].

1. United States: defines rare disease as affecting <200,000 patients (<6.37 in 10,000 based on US population of 314 million).
2. EU: defines rare disease as affecting <5 in 10,000 (<250,000 patients, based on EU population of 514 million).
3. Japan: defines rare disease as affecting <50,000 patients (<4 in 10,000 based on Japan population of 128 million).

There are around 7,000 rare diseases with an estimated 30 million Americans affected with rare disease in the US. These numbers suggest that even though the individual disease may appear to be rare in the population as a whole, the total number of people affected with orphan diseases is large considering the number of rare diseases. Development of treatments for orphan diseases was not lucrative for the pharmaceutical industry until recently considering the low population affected by each disease. Such a low population almost guaranteed that the drug developer could not recover its drug development costs. Therefore, recognizing the economic barriers to the development of rare disease therapies, the United States became the first to incentivize the process through the Orphan Drug Act of 1983. This was later adopted by Japan in 1993 and by the EU in 2000.

An orphan designation status is granted to an orphan drug if the developer provides evidence that the product is reasonably efficacious in treating a particular rare disease. Such a designation status makes the drug developer eligible for various financial incentives under the Orphan Drug Act depending upon the country or region. These are given below:

United States:

- 7 years of marketing exclusivity from approval
- 50% tax credit on R&D cost
- R&D grants for Phase I to Phase III clinical trials
- Waiver on user fees.

EU offers 10 years of marketing exclusivity from approval.

The introduction of the Orphan Drug Act in United States and the EU has provided a steady boost to the development of rare disease therapies since 1983. Since the introduction of the legislation, the number of orphan drug designations has been steadily increasing, with US recording a maximum of around 350 designations in 2015 [1] (Figure 9.1).

Lower research and development (R&D) costs, pricing incentives, quicker approval timelines, and a longer patent life for the product has attracted many pharmaceutical giants in the rare-disease field. Sales for orphan drugs are expected to almost double between 2016 and 2022 and reach $209 billion with a CAGR of >11.1%. Orphan drugs are estimated to form 21.4% of worldwide prescription sales by 2022 [1] (Figure 9.2).

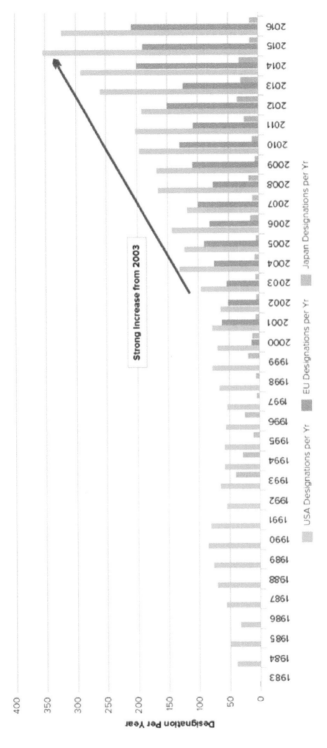

Figure 9.1 *US, EU, and Japan orphan designations per year (1983–2016). (From Evaluate pharma orphan drug report, evaluate pharma, 5th edition, May 2018.)*

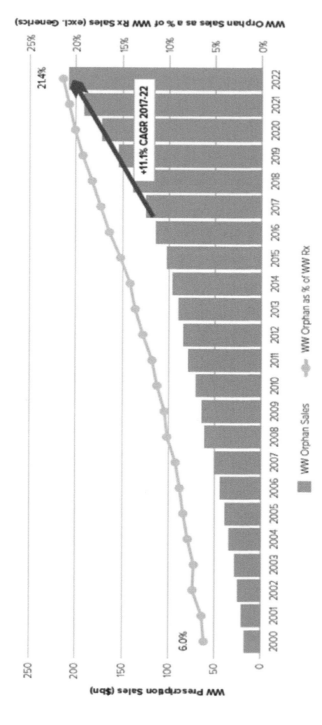

Figure 9.2 *Worldwide (WW) orphan drug sales and share of prescription drug market (2000–2012). (From Evaluate pharma orphan drug report, evaluate pharma, 5th edition, May 2018).*

9.1 Why rare diseases and orphan drugs are of US national interest

Most rare diseases are often severely disabling and cause substantial reduction in quality of life and affect an individual's potential for education and earning capabilities. Rare diseases pose a considerable burden on the affected families who face loss of social and economic opportunities along with lack of expert medical care. Most patients with rare diseases face diagnostic delays due to incorrect diagnosis that leads to futile medical interventions. Adequate drugs for many rare diseases have not been developed since a pharmaceutical company that develops an orphan drug generates relatively small sales in comparison to the cost of developing the drug and fears financial loss. Therefore, in the national interest of United States and its 30 million population afflicted with rare diseases, it is necessary to develop newer drugs and therapies for rare/orphan diseases, provide rare disease patients with better treatment and medical care.

Such an initiative would improve their quality of life, make them productive, and decrease socio-economic implications on the society. Recognition that rare diseases are an important medical and social issue in the American society is constantly growing [2].

United States has recognized the need for developing newer therapies for patients with rare diseases and has signed into law the Orphan Drug Act (ODA) that grants a special status to a drug or a biologic product to treat a rare disease or condition upon request of a sponsor. This status is referred to as "orphan designation." Such a drug must meet certain criteria specified in the ODA and FDA's regulations at 21 CFR Part 316 [3].

A sponsor whose drug is granted "Orphan Designation "status is eligible for various incentives by the Office of Orphan Products Development (OOPD)-FDA. Drug products that are approved by the FDA to treat rare/orphan diseases are eligible for a seven-year marketing exclusivity with tax credits up to 50% of the clinical development costs. The sponsor also receives assistance from OOPD-FDA while developing the drug. An Orphan Products Grant Program by the OOPD-FDA also funds 10–20 new grants per year for rare diseases. Fifty percent of these grants are awarded to company-academic institution collaborations [4].

The ODA thus offers a major avenue of investment for drug developing companies.

9.2 What are botanical drug products?

Botanical Drug Products as per USFDA are as follows:

Botanical products are finished, labeled products that contain vegetable matter as ingredients.

A botanical product may be a food (including a dietary supplement), a drug (including a biological drug), a medical device (e.g., gutta-percha), or a cosmetic under the Act. An article is generally a food if it is used for food (21 U.S.C. 312(f)(1)). Whether an article is a drug, medical device, or cosmetic under the Act turns on its "intended use" (21 U.S.C. 312(g)(1)(B) and (C), (h) (2) and (3), (i)). "Intended use" is created by claims made by or on behalf of a manufacturer or distributor of the article to prospective purchasers, such as in advertising, labeling, or oral statements.

The term botanicals include plant materials, algae, macroscopic fungi, and combinations thereof. It does not include:

- Materials derived from genetically modified botanical species (i.e., by recombinant DNA technology or cloning).
- Fermentation products (i.e., products produced by fermentation of yeast, bacteria, and other microscopic organisms, including when plants are used as a substrate, and products produced by fermentation of plant cells), even if such products are previously approved for drug use or accepted for food use in the United States (e.g., antibiotics, amino acids, and vitamins).
- Highly purified substances (e.g., paclitaxel) or chemically modified substances (e.g., estrogens synthesized from yam extracts) derived from botanical sources.

CDER guidance addresses all botanical drug products (in all dosage forms) that are regulated under the Act, except those also regulated under section 351 of the Public Health Service Act (42 U.S.C. 262). Although this guidance does not address drugs that contain animals or animal parts (e.g., insects, annelids, shark cartilage) and/or minerals, either alone or in combination with botanicals, many scientific principles described in this guidance may also apply to those products. When a drug product contains botanical ingredients in combination with either (1) a synthetic or highly purified drug or (2) a biotechnology derived or other naturally derived drug, this guidance only applies to the botanical portion of the product [5].

9.3 Botanical drugs candidates as "orphan drugs"

The following are some of the companies that are developing botanical drug products (BDPs) targeting in some of the rare diseases.

- Green Cross Holdings Corp released Shinbaro, a botanical drug for degenerative arthritis, which is increasing due to poor living habits and aging. The company is expanding its business into the field of musculoskeletal disorders [6]. Degenerative Arthritis is one of the rare diseases (6).
- Polyphenon Pharma, an emerging research-based pharmaceutical company, announced that the FDA has granted orphan drug designation to its botanical drug, Polyphenon E(R), for the treatment of chronic lymphocytic leukemia (CLL). A Phase II study is currently underway at the Mayo Clinic in Rochester, Minnesota where researchers are studying the effects of an oral daily dose of Polyphenon E in CLL patients [7].
- GW Pharmaceuticals plc (AIM: GWP, Nasdaq: GWPH, "GW") have announced that the FDA has granted orphan drug designation for Epidiolex®, product candidate that contains plant-derived cannabidiol (CBD) as its active ingredient, for use in treating children with Dravet syndrome, a rare and severe form of infantile-onset, genetic, drug-resistant epilepsy syndrome [8].
- Regenera Pharma is conducting a randomized SAD and MAD study evaluating the safety and tolerability of RPh201 (botanical drug product) in healthy subjects and in adults with Alzheimer's disease [9].

To augment further research and development in Botanical Drug Products, NIH has also taken following initiative:

Studies of the safety, effectiveness, and biological action of botanical products are major focuses for the five dietary supplement research centers selected to be jointly funded by the Office of Dietary Supplements (ODS) and the National Center for Complementary and Alternative Medicine (NCCAM), two components of the National Institutes of Health. The NIH's National Cancer Institute is co-supporting two of the five centers.

The competitive awards, approximately $1.5 million each per year for five years, were made to Pennington Biomedical Research Center, Baton Rouge, La.; University of Illinois at Chicago; University of Illinois at Urbana-Champaign; University of Missouri, Columbia; and Wake Forest University Health Sciences, Winston-Salem, N.C.

These five interdisciplinary and collaborative dietary supplement centers, known as the Botanical Research Centers (BRC) Program, are expected to advance understanding of how botanicals may affect human health. "Eventually, the program may provide data that translates to new ways to reduce disease risk," explained Paul M. Coates, Ph. D., director of ODS. "Until then, the research from these centers will help the public make informed decisions about botanical dietary supplements" [10].

9.4 Process for botanical drug product development for rare diseases

9.4.1 Chemistry, manufacturing, and controls of botanical drug products

9.4.1.1 Objectives

- To identify differences in requirements for botanical drug products compared with the requirements for new drugs based on FDA's new chemistry guidance document botanical drug products, June 2004.
- To identify areas where additional information will be required to support both Phase III trials, and ultimately the submission and approval of an NDA.

To determine if botanical product meets the definition of a drug based on "intended use" under the FD&C Act.

For clinical studies, determine phase of investigation IND is intended to support and amount of CMC information that will be needed as shown in following Table 9.1.

Determine if product meets combination drug regulations

- Not considered fixed combination drugs if from single part of plant (leaves, stems, roots or seeds) or from a single species of algae or macroscopic fungi.
- Considered combination drug if composed of multiple parts of a single species or of parts of different species of plant, algae, or macroscopic fungi.
- For marketing, determine if, product meets or can meet OTC monograph will require approved NDA or ANDA.

9.4.1.2 Evaluation

The extent of CMC information required for an IND for a botanical drug product depends on information already known and the phase of clinical investigation the IND is intended to support as follows (Table 9.2):

Table 9.1 Preliminary Evaluations for Botanical Drug Product

Parameter	Issue
Novelty of drug	Extent previously studied
Safety	Drug product's known or suspected risks
Marketing status	Whether legally marketed in the US under DSHEA; whether legally marketed outside US
Marketing history development phase of drug	Data requirements may vary for preclinical studies

Table 9.2 CMC Information Required for Non-marketed Botanical Products and Products with Known Safety Concerns

Material	Information Required
Botanical Raw Material	Identification by trained personnel
	Certificate of authenticity
	List of all growers and/or suppliers
Botanical Drug Substance	Qualitative and quantitative description
	Name and address of manufacturer
	Description of manufacturing process
	Quality control tests performed
	Description of container/closure system
	Available stability data
	Container label
Botanical Drug Product	Qualitative description of finished product; composition of finished product
	Name and address of manufacturer
	Description of manufacturing process
	List of quality control tests performed
	Description of container/closure system
	Available stability data
	Placebo
	Labeling
	Environmental assessment or claim of categorical exclusion

- Legally marketed in US under DSHEA
- Phase I, II, and III studies to be conducted
- Novel drug
- Legally marketed outside US
- Botanical product has known or suspected risks

Key differences in requirements for botanical drugs as compared to other new drugs include:

- Not essential to identify active constituent(s)
- FDA may rely on combination of tests and controls to ensure identity, purity, quality, strength, potency and consistency of drug substance and drug product testing

Key FDA draft recommendations could be as follows (Tables 9.3 and 9.4):

- Use a consistent formulation for both drug substance and dosage form throughout the clinical trials, unless this proves impossible.
- Determine the parameters and specifications for batch or lot release testing as the clinical studies progress.

- Retain sufficient quantities of the botanical raw material and drug substance from the same batch for future chemical and physical characterization.
- Include historical and scientific information (and its source) or prior human use of the botanical product and each of its ingredients in traditional foods and drugs.
- Identify the harvest location, growth conditions, stage of plant growth at harvest time, collection/washing/drying and preservation procedures, and handling, transportation, and procedures.
- Provide quantity and sequence of addition, mixing, grinding or extraction if more than one botanical raw material is introduced to produce a multi-herb substance.

Table 9.3 Target Rare Diseases Therapeutic Areas wherein Botanical Drugs Can Be Developed as "Orphan Drugs" through Reverse Pharmacology Approach with the Help of Ayurveda, Indian System of Medicine

Sr. No.	Rare Disease
1	AIDS (Acquired Immune Deficiency Syndrome)
2	Alopecia Areata
3	Alzheimer's Disease
4	Anemia of Chronic Disease
5	Arthritis, Psoriatic
6	Benign Essential Blepharospasm
7	Burning Mouth Syndrome
8	Candidiasis
9	Chronic Inflammatory Demyelinating Polyneuropathy
10	Crohn's Disease
11	Diverticulosis
12	Elephantiasis
13	Familial Hypercholesterolemia
14	Fibromyalgia

(Continued)

Table 9.3 (*Continued*) Target Rare Diseases Therapeutic Areas wherein Botanical Drugs Can Be Developed as "Orphan Drugs" through Reverse Pharmacology Approach with the Help of Ayurveda, Indian System of Medicine

Sr. No.	Rare Disease
15	Gastritis, Chronic, Erosive
16	Heavy Metal Poisoning
17	Hepatic Encephalopathy
18	Hiccups, Chronic
19	Hodgkin's Disease
20	Huntington's Disease
21	Hyperemesis Gravidarum
22	Hypothyroidism
23	Irritable Bowel Syndrome
24	Lichen Planus
25	Lupus
26	Malaria
27	Parkinson's Disease
28	Prostatitis
29	Psoriasis
30	Radiation Sickness
31	Renal Glycosuria
32	Rheumatic Fever
33	Rickets, Vitamin D Deficiency
34	Tinnitus
35	Tuberculosis
36	Typhoid
37	Ulcerative Colitis
38	Urticaria, Cholinergic
39	Urticaria, Cold
40	Urticaria, Papular
41	Urticaria, Physical
42	Vitiligo

Table 9.4 Strategy Steps Involved in Botanical Drug Product Development as Orphan Drugs

BDP Development Process	Influence On	Subject Knowledge Requirement
Disease Epidemiology Research & Market Research	Market & Price Determination	Clinical Medicine
		Preventive Social Medicine, Pharmacology
Pharmacoeconomics of Target Disease		Marketing Management & Economics
Market Research for Target Disease		Marketing Management
Product Niche for Target Disease		Marketing Management, Indian System of Medicine, Evidence-Based Medicine, Research Methodology, Biotechnology Management
Reverse Pharmacology for Target Disease and Leads Database	Drug Development Cost Optimization	Indian System of Medicine, Evidence-Based Medicine, Research Methodology, Biotechnology Management
Prospective Lead for Botanical Drug Product Candidate		
Was the Prospective Lead Marketed in USA?	CMC Development Cost	Marketing Management
Was the prospective lead marketed outside USA?		Pharmacology Biotechnology Management
Safety, Toxicity, Human Use Secondary Data Research and BDP Lead	CMC/NDA/IND filing Budget	Pharmacology, Research Methodology, Biotechnology Management
Predetermine the Intended Use of BDP Lead	Clinical Trials Phase I, IIa, IIb, IIIa, IIIb Budget	Clinical Medicine, Clinical Pharmacology, Evidence-Based Medicine, Translational Research, Biotechnology Management
Target Disease Primary and Secondary End Point/Indication Determination		
Phase Arms	IIIa, IIIb Clinical Phase Trial Budget	
Chemistry, Manufacturing, Control		
BDP Novelty Information	NDA/IND Filing Budget	Research Methodology
BDP Safety Information		Clinical Pharmacology
BDP Market Status Information		Marketing Management, Applied Economics
BDP Market History Information		
BDP Known-Risk Information		Research Methodology, Clinical Pharmacology
BDP Meets OTC Monograph or NDA or ANDA will Require?	Product Placement in Market	Marketing Management, Applied Economics
Is Market Exclusivity Desired?		
BDP Good Agricultural Practices Protocols	Product Cost	Systems and Operations Management, Applied Economics
Bioassays, *in Vivo* assays, Microarray, Fractional Isolation	Cost vs Benefit Ratio	Pharmaceuticals Engineering, Biotechnology Management
Soluble Marker Identification	Final Product Cost	
BDP Lab Scale and Pilot Scale Process		

(Continued)

Table 9.4 (*Continued*) Strategy Steps Involved in Botanical Drug Product
Development as Orphan Drugs

BDP Development Process	Influence On	Subject Knowledge Requirement
Pre Clinical Studies (Safety & Toxicity)		
In Vitro/ex Vitro Study	Phase I Clinical Trial Budget	Clinical Medicine, Clinical Pharmacology, Evidence-Based Medicine, Translational Research, Biotechnology Management
Animal Toxicity/Safety (Acute/Chronic)		
Carcinogenecity/Reproduction/Teratology Study		
Preferred Mode of Drug Delivery		
Compound Patent Filing		
BDP Manufacturing Determination		
BDP Pilot Scale-Up Batch	Final Product Cost Estimation	Pharmaceuticals Engineering, Biotechnology Management
BDP Description of Manufacturing Process		
BDP QA/QC, VPs, SOPs		
Container/Closure System Description		
Available Stability Data		
Placebo	Phase III Clinical Trial Budget	Clinical Medicine, Clinical Pharmacology, Evidence-Based Medicine, Translational Research, Biotechnology Management
Labelling (Official Instructions to Use)		
Environmental Assessment, Categorical Claim Exclusion	Final Product Cost Estimation	
NDA Filing with Sufficient Safety and Efficacy Evidence		
Labelling Review	FDA Process	Regulatory Affairs
If Not Satisfactory, Pending Results		
Sponsor Revision		
Manufacturing Site Inspection		
Clinical Trial Site Inspection		
If Not Satisfactory, Pending Results		
Sponsor Review with All Additional Inputs		
NDA Action		
NDA Approved		
Drug Marketing		Clinical Medicine, Clinical Pharmacology, Evidence-Based Medicine, Translational Research, Biotechnology Management, Marketing Management
Drug Promotion		
Post-Market Surveillance Plan		
NDA Not Approved		
INDA Filing		

(*Continued*)

Table 9.4 (*Continued*) Strategy Steps Involved in Botanical Drug Product Development as Orphan Drugs

BDP Development Process	Influence On	Subject Knowledge Requirement
If BDP was Marketed Outside USA, and or in USA With No Known Safety Issues:		
Additional Information on CMC and Non-Clinical Safety	Final Product Cost Estimation	Clinical Medicine, Clinical Pharmacology, Evidence-Based Medicine, Translational Research, Biotechnology Management, Marketing Management, Pharmaceutical Engineering
Clinical Trials I, II, III Data		
Process Patent Filing		
If BDP was Marketed Only Outside USA, and With Known Safety Issues:		
CMC Data	Final Product Cost Estimation	Clinical Medicine, Clinical Pharmacology, Evidence-Based Medicine, Translational Research, Biotechnology Management, Marketing Management, Pharmaceutical Engineering
Clinical Trials I, II, III Data		
Process Patent Filing		
If BDP Was Not Marketed Inside & Outside USA, and with Known Safety Issues:		
Is BDP in Traditional Use? Yes,		
CMC Data	Final Product Cost Estimation	Clinical Medicine, Clinical Pharmacology, Evidence-Based Medicine, Translational Research, Biotechnology Management, Marketing Management, Pharmaceutical Engineering
Clinical Trials I, II, III Data		
Process Patent Filing		
Is BDP in Traditional Use? No,		Ethno Pharmacology, Evidence-Based Medicine, Botanical Drug Products, Research Methodology, Biotechnology Management
Non Clinical Safety Data	Final Product Cost Estimation	Clinical Medicine, Clinical Pharmacology, Evidence-Based Medicine, Translational Research, Biotechnology Management, Marketing Management, Pharmaceutical Engineering
NDA Filing with Sufficient Safety and Efficacy Evidence	FDA Process	Regulatory Affairs
NDA Review by FDA CDER		
Proposed Monograph or Proposed Amendment		
Final Monograph and Final Amendment		
Final Monograph and Final Amendment Published in Federal Register and Code of Federal Regulations		
Drug Marketing	Product Pricing Policy	Clinical Medicine, Clinical Pharmacology, Evidence-Based Medicine, Translational Research, Biotechnology Management, Marketing Management, Pharmaceutical Engineering
Drug Promotion		
Post-Market Surveillance Plan		

9.5 Conclusion

Rare diseases are attracting interest worldwide by pharmaceutical industry to develop new orphan drugs. Industry might get regulatory advantage in various territories for its product market placement, as rare disease in one country might be potential disease in another country, for example AIDS. Botanical drug products due to their poly composite nature can have scientific advantage in terms of their safety (history of use), efficacy establishment through reverse pharmacology (long history of use in some other similar conditions), and can be cheaper comparatively in drug development to any other new drug molecules.

References

1. Evaluate pharma orphan drug report, evaluate pharma, 5th edition, May 2018.
2. A. Schieppati, J.-I. Henter, E. Daina and A. Aperia, Why rare diseases are an important medical and social issue, *Lancet* 371 (2008), 2039–2041.
3. Designating an orphan product: drugs and biological products. Available at https://www.fda.gov/ForIndustry/DevelopingProductsforRareDiseasesConditions/HowtoapplyforOrphanProductDesignation/default.htm. Accessed August 8, 2018.
4. M.T. Thomas, The orphan drug act and the development of products for rare diseases. US Food and Drug Administration. Available at https://www.fda.gov.
5. US-FDA (CDER), Botanical drug development guidance for industry, (2016), 34.
6. GC Pharma. Available at https://www.globalgreencross.com/product/list_detail?p_idx=35. Accessed August 8, 2018.
7. Polyphenon pharma receives orphan drug designation for its botanical drug, Polyphenon E, for the treatment of CLL. Available at https://www.medicalnewstoday.com/releases/116391.php. Accessed August 8, 2018.
8. GW pharmaceuticals provides update on orphan program in childhood epilepsy for Epidiolex®. Available at https://www.gwpharm.com/about-us/news/gw-pharmaceuticals-provides-update-orphan-program-childhood-epilepsy-epidiolex%C2%AE. Accessed August 8, 2018.
9. A randomized SAD and MAD study evaluating the safety and tolerability of RPh201 in healthy subjects and in adults with alzheimer's disease—Full text view—ClinicalTrials.gov. Available at https://clinicaltrials.gov/ct2/show/NCT01513967. Accessed August 8, 2018.
10. NIH Announces five botanical research centers. Available at https://nccih.nih.gov/news/2010/083110.htm. Accessed August 8, 2018.

Botanical Drug Products Using Advanced Drug Delivery Systems
Recent Trends

Param Patel, Anas Hanini, Achal Shah, Adam Mohamed,
Kareem Elgendi, and Yashwant Pathak

Contents

10.1 Introduction

There have been a multitude of ways to treat human health throughout the years. One of those ways is through the use of botanical drug products. Botanical drugs are prepared in conjunction with numerous types of herbs that can be used as ingredients[1]. The conjunction of herbs must be prepared in special ratios so that they can be administered for proper use [2]. Human diseases and their treatments can be done through botanical drug products intended for treatment, prevention, and diagnosis. Vegetable and plant extracts, fungi, or algae can serve as vital materials in the use of botanical drug products. The administration of botanical drug products can vary in

Table 10.1 Medicinal Botanicals with Thier Respective Therapeutic Uses

Herbal Drug	Scientific Name	Treatment
Milk thistle	*Silybum marianum*	Type II diabetes
Saw palmetto	*Serenoa repens*	Chronic abacterial prostatitis
Garlic	*Allium sativum*	Hepatopulmonary syndrome
Valerian	*Valeriana officinalis*	Coronary heart disease
Kava	*Piper methysticum*	Anxiolytic antidepressant activity
Chamomile	*Matricaria recutita*	Generalized anxiety disorder
Echinacea	*Echinacea purpurea*	Common cold
Curcumin	*Curcuma longa*	Healing of peptic ulcer
Ginseng	*Panax quinquefolius* or American ginseng	Reduction of postprandial glycaemia and improvement from cancer related fatigue
Ginger	*Zingiber officinale*	Nausea from pregnancy and osteoarthritis (OA) in the knee
Ginkgo	*Ginkgo biloba*	Peripheral arterial insufficiency Raynaud's disease
St. John's wort	*Hypericum perforatum*	Peptic ulcer and chronic gastritis Major depressive disorder
Essential oils from Atractylodes	*Atractylodes lancea*	Anti-arthritis
Green peppers	*Capsicum annuum*	Anti-herpes

a number of ways. Some of those methods of administration may be, but not limited to, injections, powders, solutions, tablets, topicals, and capsules. With recent advancements in botanical drug development, there have been more precautions taken into consideration with the quality control of botanical sources [3] (Tables 10.1 and 10.2).

Drug delivery is the administration of various pharmaceutical substances with the aim of acquiring a therapeutic effect in mammals [4]. Drug delivery is usually one of the main obstacles faced when delivering a treatment. However, through extensive research on the development of smart drug delivery, there has been a significant increase in methods to administer various drugs. The progress has been remarkable and has allowed researchers to better understand the limitations and advantages in administration. Some of these advantages are nanotechnology, as well as nanoparticle delivery through transmission methods such as liposomes and niosomes (Table 10.3).

Table 10.2 Phytochemicals Derived from Medicinal Plants with Chemical Structure

Compounds	Chemical Structure	
Curcumin		 *Curcuma longa*
Lycopene		 Gac fruit (*Momordica cochinchinesis*)
Tetrandrine		 *Stephania tetandra*
Tanshinone		 *Salvia miltiorrhiza*
Silibinin		 *Silybum marianum*

Table 10.3 Nano Carriers and Constituents

Type of Nanocarrier	Definition	Nanomaterial Used
Liposomes	Small artificial vesicles of spherical shape that can be created from cholesterol and natural non-toxic phospholipids	Phosphatidylcholine Phosphatidylglycerol Phosphatidylethanolamine Phosphatidylserine
Polymeric micelles	Polymeric micelles are self-assembling nanoscopic with a hydrophobic core and hydrophilic shell structure formed by amphiphilic block copolymers [5]	Poly(ethylene oxide)-poly(benzyl-L-aspartate) Poly(N-sopropylacrylamide)-polystyrene
Dendrimers	Three-dimensional structured polymer containing highly ordered oligomeric and polymeric compounds which are formed by repeating sequences of ammonia or pentaerythritol known as 'initiator cores' [6]	Polyamidoamine (PAMAM)
Polymeric nanoparticles	Polymeric nanoparticles include nanospheres and nanocapsules and are small particles that are 10–10,000 nm in size. They are able to alter drug activity, control time of release, and increase the adhesiveness of the drug during entry into the skin [7]	Poly-(ε-caprolactone) (PCL) Poly(lactide-coglycolide) (PLGA) Chitosan Alginate Albumin Gelatin
Solid-lipid nanoparticles	Nanostructures made from solid lipids	Glyceryl behenate Stearic triglyceride Cetyl palmitate Glycerol tripalmitate

10.2 Applications of advanced drug delivery for botanical drug products

Numerous advanced drug delivery methods are available to administer botanical drug products. Different methods are used all around the world. For certain methods to work, the drug must be encapsulated into the nanocarriers (liposomes, solid lipid, polymeric nanoparticles, dendrimers, micelles, niosomes) [8]. The bulk material property can be used to synthesize different nanomaterials such as, polymeric nanoparticles, lipid based nanoparticles, and metal nanomaterials [9–11]. Pharmacological studies have investigated the chemical composition of plant extracts for centuries to identify their various therapeutic uses. Advancements in the research of biological, pre-clinical, and biochemical studies have shown the scientific usefulness of botanical products for various treatments such as carcinomas, skin disorders [12], rheumatological disorders [13], asthma and allergies [14], liver disease [15], etc. Botanical products pose major challenges in drug delivery due to their chemical structure, hydrophilic nature, and poor bioavailability in regions of the human body. In order for herbal products to be effective, it is dependent upon the levels of the active compounds administered. The use of advanced drug delivery helps overcome these challenges [12].

Advanced drug delivery methods include the use of nanoparticles as mentioned above. For treatment of human diseases, the use of pulmonary and nasal drugs is on the rise, causing for a higher demand for nanoparticles to facilitate the dispersion and administration of the drugs. Liposomes, proliposomes, microspheres, gels, prodrugs, cyclodextrins are all forms of nanoparticles for capsule delivery. Depending on their composition, it can control the time when it starts to degrade and release the drug, and the duration for when the drug is administered for. Transdermal is another popular method these drugs can be administered [4]. Furthermore, drugs are sometimes co-delivered either with other drugs or with genes by the use of many forms of nanoparticles. This helps to overcome multi-drug resistance of chemical drugs and achieve a synergistic therapeutic effect [16]. Another form of drug delivery is brought about by blending polymers degrading by hydrolysis and pluronic surfactants (block copolymers of polyethyleneoxide (PEO) and polypropyleneoxide (PPO)). The hydrophobicity and hydrophilicity can be adjusted by changing the PEO/PPO ratio. This allows for the control of phase separation with PLA. Water content of the polymer can be controlled by blending different kinds of block copolymers and by adjusting the relative ratios [39] (Tables 10.4 and 10.5).

Factors affecting drug delivery [cell 92]:

Table 10.4 Influencing Physiological Factors on Drug Delivery

Based On	Stimulus	Mode
Physical Stimuli	Osmotic pressure	Drug delivery that is mediated through water permeability.
	Hydrodynamic pressure	Such stimulus helps the drug to be released through an opening due to the hydrodynamic gradient.
	Mechanical force	Contains a mechanically activated pump. Responsible for a pressure sensitive delivery and first pass elimination.
	Sonophoresis	Drug delivery is activated through an ultrasonic energy.
	Iontophoresis	Equipped to activate and diffuse a charged drug through an electrical current.
Chemical Stimuli	Salt concentration	Mediates delivery of an ionic drug due to the preparation by ionizable drug with ion exchange resin.
	pH	Drug delivery should be through an intestinal tract and ulcer stomach through floating delivery.
	Hydrolysis	Activates the release of a drug molecule through hydrolysis-induced degradation of polymer chains.
Biochemical Stimuli	Enzyme	Enzymatic hydrolysis of polymers initiates the drug delivery due to polymer chains being fabricated with biopolymers.
	Biochemical	A concept of enzyme-activated, biodegradation feedback regulated delivery has been equipped.
Environmental Stimuli	Temperature	Change in temperature leads to the shift in balance of hydrophilicity and hydrophobicity.
	Light	Photons help polymers undergo isothermal phase transition.

Table 10.5 Drug Delivery Carrier Compounds

Sources	Materials
Natural	Gelatin, albumin, alginate, collagen, chitosan
Synthetic	Poly(lactide) (PLA), poly(lactideco-glycolide) (PLGA), poly-ε-caprolactone (PCL), Polymethyl metha acrylate
Metallic	Iron oxide, gold, silver, gadolinium and nickel
Ceramic	Silica, alumina and titani

10.3 Polymeric nanoparticles for drug delivery

The use of polymers has gained much popularity due to the unique bio-logical characteristics they possess. Having a wide array of structures [16], simple synthesis, biological compatibility, biological degradability [17], aqueous solubility [18], and controllable molecular weight [19]. Nanoparticles have also been revolutionary for the use of smart drug delivery. Nanoparticles range from 10 to 1000 nm in size [20]. The drug is either attached or encapsulated into the nanoparticle matrix, and depending on the preparation, various nanoparticles such as nanocapsules and nanospheres can be obtained [20–21]. Polymeric nanoparticles are made by natural and synthetic polymers. The research on using polymers as therapeutic agents has shed new light on treatments to administer polymeric nanoparticles for smart drug delivery to target specific sites [16–17]. Polymeric nanoparticles have properties allowing scientists to alter the encapsulated drug being administered and control its time of release [22]. The capability for nanoparticles to reach its target site is influenced by hydrophobicity, charge, and size of the particle. The size of nanoparticles influences the interactions between the nanoparticle and its ability to penetrate the cell membrane barrier [19]. This is an important factor because only small selective particles can diffuse across the cell membrane. Polymeric nanoparticles consist of multiple polymer chains that are either hydrophilic or hydrophobic and self-assemble in an aqueous environment. The hydrophobic region forms the inner side of the polymers' membrane, facing away from the aqueous environment, whereas the hydrophilic region forms the outside of the polymers' membrane facing towards the aqueous environment [23]. The advantage of hydrophobicity in drug-loaded polymeric nanoparticles provides a controlled release of drugs over a period of time which reduces unwanted side effects [16,23]. A number of drugs can be administered via different routes using nanoparticle carriers. Target sites of nanoparticles may be (but are not limited to) the brain, lungs, liver, lymphatic system, spleen, arterial walls, or systemic circulation [9]. In comparison to conventional drug delivery methods, patients will be more compliant because of the advantages polymeric nanoparticles have, such as dosage frequency. Other advantages include permeability and solubility of botanical drug products, and the ability to increase bioavailability.

The future of nanoparticle drug delivery will help overcome more challenges of certain botanical drugs which dissolve poorly in aqueous environments and are less permeable [19]. They have been used to (potentially) treat cancer as well as deliver vaccines in a less invasive and more effective manner [9]. Polymeric drug delivery systems also are very versatile. They have controlled release drug delivery systems, activation modulated drug delivery, environmental activation/stimuli responsive smart delivery systems, dual and multi-stimuli responsive smart delivery systems, feedback regulated drug delivery systems, and site targeting drug delivery systems. There can also be any mix of the previous attributes in polymeric nanoparticles [16] (Table 10.6).

Table 10.6 Nano Drug Delivery Forms and Manufacturing Methods

Nanostructure	Method
Liposomes	Sonication
	Electroformation
	Extrusion from diluted lamellar dispersions
	High-shear homogenization
	Reverse-phase evaporation
	Gel exclusion chromatography
	Freeze lyophilization
	Calcium-induced fusion
	Detergent dialysis
	Ultracentrifugation
Polymeric micelles	Dialysis
	Solution-casting
	Direct dissolution
Dendrimers	Tomalia's divergent growth approach
	Convergent growth approach
	Orthogonal coupling strategy
Polymeric nanoparticles	Ionic gelation
	Coacervation
	Solvent evaporation
	Spontaneous emulsification/solvent diffusion
	Salting out/emulsification-diffusion
	Supercritical fluid technology
	Polymerization

(Continued)

Table 10.6 (*Continued*) Nano Drug Delivery Forms and Manufacturing Methods

Nanostructure	Method
Solid-lipid nanoparticles	High shear homogenization
	Ultrasound dispersion technique
	High pressure homogenization
	Solvent emulsification/evaporation microemulsion and solvent diffusion
Nanocapsules	Microemulsion
	Miniemulsion polymerization and interfacial polymerization
Ceramic nanoparticles	Template synthesis
	Hot pressing technique
	Controlled hydrolysis in micellar medium
Metallic nanoparticles	Gas phase deposition
	Electron beam lithography
Nanoemulsions	Spontaneous emulsification
	High pressure
	Ultrasonic homogenization

10.4 Liposomes to deliver botanical drug products

Liposomes are a class of nanoparticle carriers that have shown substantial usefulness in many drug delivery applications. These drug carrier vesicles are produced from lipids and offer protection to the drug contained inside of its matrix [9]. Liposomes are a phospholipid nanoparticle carrier system made of a bilayer membrane and take form in the shape of a bubble. Liposomes have high biocompatibility, can be used in conjunction with a wide array of drugs, have a simple preparation process, and are easily manipulated *in vivo* [18]. Properties of liposomes can vary due to surface charge, lipid composition, size, and preparation methods. The components that make up the bilayer determine its permeability, fluidity, rigidity, and charge [24]. Conventional liposomes can have single or multi bilayers that constitute from cholesterol and phospholipids similar to human cells. The fluidity and rigidity of the bilayer depend on the phospholipid to cholesterol ratio [25]. The presence of cholesterol gives it the ability to tighten the lipid bilayer which is important so that the encapsulated drug does not leak out. The liposomes membrane ability to switch from a fluid phase to a solid phase reduces the leakage of the encapsulated drug. Research also indicates that the incorporation of sphingomyelin into the bilayer can influence the fluidity and rigidity to reduce leakage. There are two main ways to trigger a liposome to release its contents. The first type of triggers are remote triggers including ultrasound, light, and heat. The second type of triggers are local triggers innate to the cellular organelles including proteins, enzymes, and pH chains [26]. It is important to mention that the desired circulation time of traditional liposomes will not always be reached. Stealth liposomes have an enhanced composition charge and size which enables them to last longer in circulation [24]. The liposome drug delivery system for delivering botanical drug products has advantages, which include bioavailability, sustained delivery, protection from degradation, protection from toxicity, solubility enhancement, pharmacological activity enhancement, and an increase in distribution to tissue macrophages. Liposomes meet two important criteria for botanical drug delivery; delivery of the active ingredient of the botanical drug to the target site and delivery of the botanical drug at a rate suitable for the body over the necessary treatment time. Liposomes are nontoxic and biodegradable, encapsulating hydrophobic and hydrophilic materials. Liposomes are a great nanosized drug delivery system for botanical drugs and have a promising future for overcoming obstacles for the use of plant medicine [27].

Liposomes can vary in size and in lamilarity. They can be either unilamellar and contain a single phospholipid bilayer or they can be multilamellar which are composed of multiple liposomal membranes. The method used to assemble them can help control the physical characteristic of the liposome [24]. Unilamellar liposomes can be classified into two categories, such as small unilamellar vesicles and large unilamellar vesicles. These different classes of liposomes usually require different methods of preparation. The most common method for preparation of multilamellar liposomes is using the thin film hydration procedure. In this procedure a thin film of lipid is hydrated in an aqueous solution/buffer at temperatures higher than the transition temperature of lipids. The desired drug to be encapsulated into the liposome is placed either into the lipid film or the aqueous hydration buffer. The reduction of the liposome size can affect the amount of drug that can be encapsulated within it. Other methods for preparation have been developed for large unilamellar vesicles which include but are not limited to solvent injections, calcium induced fusion, detergent dialysis, and reverse phase evaporation techniques. Small unilamellar vesicles can be prepared from multilamellar vesicles or large unilamellar vesicles by sonication or extrusion [40]. Spectroscopic techniques or electron microscopy measure the lamellarity of liposomes. The nuclear magnetic resonance spectrum is frequently recorded with and without paramagnetic agents causing a shift of the signals or bleaches the signal of nuclei on the outer surface of the liposome. By encapsulating a hydrophilic marker researchers can measure encapsulation efficiency [41] (Figure 10.1).

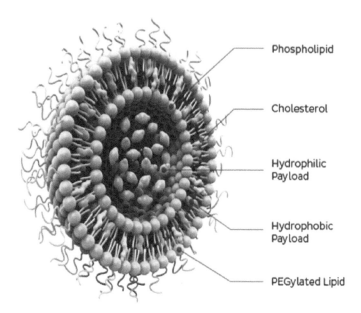

Phospholipid

Cholesterol

Hydrophilic
Payload

Hydrophobic
Payload

PEGylated Lipid

Figure 10.1 *Liposome structure.*

10.5 Phytosomes

The term "phyto" stands for plant, and the term "some" stands for cell-like. Phytosomes are a new form of drug carrier that can be used to deliver botanical extracts. The botanical extracts are bound by lipids, are water soluble, and have better bioavailability. Due to their improved pharmacokinetic and pharmacological properties, phytosomes are used therapeutically and can treat a number of diseases. Phytosomes are not liposomes due to their distinctly different structures. Phytosomes consist of a few molecules bonded together. Phytosomes have greater clinical benefits and assure the delivery of the encapsulated drug to the target tissue. They have the ability to permeate the botanical extract and allow for better absorption in the intestine. The formation of phytosomes is a safe method and its components are approved for cosmetics and pharmaceutics. Phytosomes are a cost effective form of drug delivery and are efficient when encapsulating the desired drug [42] (Figure 10.2).

Phytosomes are a cross between natural phospholipids and natural products. The formation of phytosomes also produces small compartments which protect the encapsulated botanical product from destruction from

Liposome

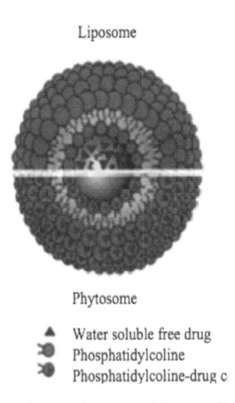

Phytosome

▲ Water soluble free drug
🌑 Phosphatidylcoline
🌑 Phosphatidylcoline-drug c

Figure 10.2 *Difference between phytosome and liposome. (From Pathak, Y.W. and Tran, H.T., Adv. Nanotechnol. Appl., 2, 121–136, 2010.)*

digestive juices. Phytosomes are able to easily reach the target site as they are permeable to the gut and can go into circulation in the body. The preparation of phytosomes require reacting 2–3 moles of a synthetic or natural phospholipid with a mole of phytoconstituents with or without a natural mixture in aprotic solvent. To maximize the usefulness of botanical products, it is necessary to have the appropriate delivery system for optimum effectiveness. Phytosomes show great potential for delivering botanical drug products, and they offer enhanced bioavailability of hydrophilic flavonoids and other botanical extracts through the gastrointestinal tract or the skin. They have many advantages over the conventional drug delivery systems. They are simply formulated and can be sold easily commercially. There are many patents that have been approved and much more to come for the processing, formulation, and application of phytosomes. The future of phytosome applications have shifted toward plant compounds that are hydrophilic [12] (Figure 10.3) (Tables 10.7 and 10.8).

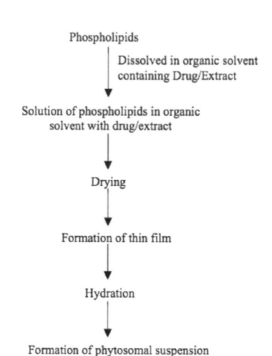

Phospholipids

| Dissolved in organic solvent containing Drug/Extract

Solution of phospholipids in organic solvent with drug/extract

Drying

Formation of thin film

Hydration

Formation of phytosomal suspension

Figure 10.3 *Phytosome preparation method. (From Patela, J. et al., Asian J. Pharm. Sci., 4, 363–371, 2009.)*

Table 10.7 Characterization Techniques for Phytosomes

Techniques	Mode
Visualization	This technique can be achieved using microscopy such as transmission electron microscopy (TEM) and scanning electron microscopy (SEM).
Vesicle size and Zeta potential	This technique can be achieved using computerized inspection system, photon correlation spectroscopy and dynamic light scattering (DLS).
Entrapment efficiency	This technique can be achieved using ultracentrifugation technique.
Transition temperature	This technique can be achieved using differential scanning calorimetry.
Surface tension active measurement	This technique can be achieved using ring method in a Du Nouy ring tensiometer.
Vesicle stability	This technique can be achieved by analyzing the size and structure of vesicles over time. The mean size is measured by dynamic light scattering (DLS) and structural changes is analyzed through transmission electron microscopy (TEM).
Drug content	This technique can be achieved using modified high-performance liquid chromatography or an appropriate spectroscopy method.
Spectroscopic evaluations	Such spectroscopic method is used to study the reciprocal interaction between the phytoconstituent and the phospholipids along with to confirm the formation of a complex.

(Continued)

Table 10.7 (Continued) Characterization Techniques for Phytosomes

Techniques	Mode
H-NMR	The NMR spectra and its constituents were discovered by Bombradelli. There is a change in signal in nonpolar solvents, without any summation of individual molecules. Flavonoids lead to a broader signal. There is broadening of all signals in phospholipids. Heating a sample leads to broader bands due to the flavonoid activity.
C-NMR	In this technique, all flavonoid carbons are invisible due to it being recorded in C6D6 at room temperature. While fatty acids retain their original sharp line shape, the signal corresponding to glycerol and choline portion of the lipid broadens.
FTIR	When phytosomes are micro dispersed in water or cosmetic gels, FTIR spectroscopy is used to control its stability.
In vitro and in vivo evaluations	Such technique is used on the basis of the expected therapeutic activity of biologically active phytoconstituents of phytosomes.

Source: Patela, J. et al., *Asian J. Pharm. Sci.*, 4, 363–371, 2009.

Table 10.8 Therapeutic Applications of Liposomes

Liposomes as drug-protein delivery vehicles	• Controlled and sustained drug release • Enhanced drug solubilisation • Altered pharmacokinetics and biodistribution • Enzyme replacement therapy and biodistribution • Enzyme replacement therapy and lysosomal storage disorders
Liposome in antimicrobial, antifungal, and antiviral therapy	• Liposomal drugs • Liposomal biological response modifiers
Liposome in tumour therapy	• Carrier of small cytotoxic molecules • Vehicle for macromolecules as cytokines or genes
Liposome in gene delivery	• Gene and antisense therapy • Genetic DNA vaccination
Liposome in immunology	• Immunoadjuvant • Immunomodulator • Immunodiagnosis
Liposomes as artificial blood surrogates	• Trauma medicine • Casualty care
Liposome as radiopharmaceutical and radiodiagnostic carriers	
Liposomes in cosmetics and dermatology	
Liposomes in enzyme immobilisation and bioreactor technology	

Source: Sharma, A. and Sharma, U.S., *Int. J. Pharm.*, 154, 123–140, 1997; Anwekar, H. et al., *Int. J. Pharm. Life Sci.*, 2, 945–951, 2011.

10.6 Polymeric micelles

Polymeric micelles are block copolymers, which are capable of self assembling into spherical structures and are promising long lasting nanomedicines currently being studied in clinical and preclinical trials [44]. Advances in the research of block copolymer synthesis have led to the development of polymeric micelles that may be used as nanoscopic drug delivery carriers. Micelles can be prepared from biodegradable and biocompatible block copolymers for advanced drug delivery. The functional groups on the surface of polymeric micelles can be used to attach pilot molecules. Current research focuses on establishing new routes for loading drugs into polymeric micelles such as chemical and physical routes [45]. Polymeric micelles have showed much potential as colloidal carriers for targeting amphiphilic drug sites and drugs that are poorly soluble in water [46]. There have been studies that illustrate the potential of polymeric micelles that can be used to target drugs actively and passively [45]. Advantages of using polymeric micelles include prolonged release, solubilisation of poorly soluble molecules, size, and protection of the encapsulated drug. In the pharmaceutical industry, polymeric micelles have been used to overcome some of the most challenging obstacles. Polymeric micelles have demonstrated the ability to improve therapeutic efficacy of botanical drugs, reduce toxicity, and efficiently deliver the drug to the desired target sites [5].

Polymeric micelles are usually composed of a few hundred block copolymers with a diameter of about 20–50 nm. The micelle contains two spherical regions; one region is a densely packed core which contains hydrophobic blocks and the second region is a shell region which consists of a dense brush of poly (ethylene oxide). Passive drug delivery can be illustrated when polymeric micelles accumulate from solid tumors upon leakage from capillaries and maintain their structure at the site. Polymeric micelles can be sterilized by filtration and require no aseptic processing. They are also used in studies pertaining to gene delivery [45]. There are several advantages that polymeric micelles have over oral administration; which is the most accepted and commonly used form of drug delivery. These advantages are its ability to withstand a longer time in the gut, protection of the encapsulated drug from the low pH environment of the GI tract, and hindrance of efflux pumps for drug uptake and accumulation [47].

Overall, polymeric micelles possess physical stability due to rapid release of encapsulated drug *in vivo*. Much research remains on the interaction of micelles with cellular components. Until recently, only a select few polymeric micelle formulations have been proven and tested as an effective form of advanced drug delivery for botanical drug products [46] (Figure 10.4).

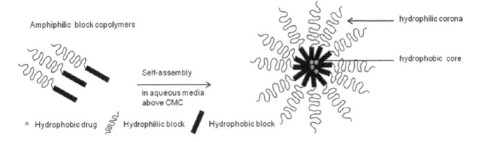

Figure 10.4 *Polymeric micelle structure. (From Jhaveri, A.M. and Torchilin, V.P., Front. Pharmacol., 5, 77, 2014.)*

10.7 Microcapsule drug delivery

Microcapsule drug carriers have become an important asset in drug delivery. Microencapsulation is a method of drug delivery in which the core material is surrounded by a membrane. This allows for better stability as well as safer handling of the drug [28]. The appropriate coating of the drug provides the necessary release characteristics of the capsule [29]. Coating should be able to become cohesive to the core material but non-reactive to it [28]. The core substance may be in the form of solids or a dispersion of liquids [29]. The properties of the core vary depending on the properties of delivery. Microcapsules can be administered through tablets, gelatin capsules, or liquid suspensions [28]. Microencapsulating a substance is explained through a variety of different reasons. Depending on the area of the body, certain drugs must be encapsulated a specific way to ensure its effectiveness. Environmental conditions, such as oxygen or heat, can impact the substance greatly if not protected by a barrier. Toxicity of substances are handled effectively through microencapsulation providing a safer drug delivery process. It also been discovered that microcapsules allow for masking of the taste in the drug. The gastrointestinal tract has been an appropriate host for microcapsule drug carriers as microencapsulation works to prevent gastric irritation [29].

The future of botanical drug delivery is dependent on various forms of administration. Microencapsulation is a very promising form of drug delivery. In a study it was shown that microcapsules with entrapped herbal water soluble extracts of *Plantago major* and *Calendula officinalis* (PCE) accelerated gastric tissue repair. Microcapsules can increase efficiency of the botanical drug, reduce the possibility of botanical extracts being destroyed by the low pH of the stomach acid, and prevent other botanical components from being metabolized by the liver prior to reaching the bloodstream. If the administered drug does not reach the minimum level necessary in the bloodstream for therapeutic effectiveness; it is known as minimum effective level [30]. A microcapsule matrix can vary in size from 1 to 300 μ. There are multiple techniques that are used to synthesize microcapsules such as polymerisation techniques, single and double emulsion techniques, solvent extraction technique, phase separation coacervation method, etc. [31].

There are also various mechanisms for how the drug is released in the system. The first mechanism involves diffusion of the matrix and entrapment of the drug to solubilise eventually leading to release. The second mechanism is surface erosion allowing the different layers to erode and eventually release the drug [31]. Different factors lead to rate of release and amount. Size, matrix, and polymer concentration are all characteristics that are taken to account for microcapsule drug release [31]. Microencapsulation has become an effective method of drug administration due to its specificity to a particular part in the body. This ensures treatment to only affected areas and avoid any harmful side effects. A study showed that an implant of Danshen by use of gelatin and chitosan as matrix polymers showed several advantages. The results showed reduced dosage administration, specificity in delivery, and taste masking [31]. This implantation through micro surgery worked to prevent inflammatory side effects (Figure 10.5).

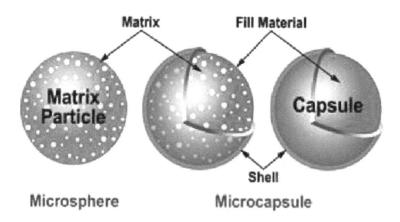

Figure 10.5 *Microcapsule and microsphere. (From Agnihotri, N. et al., Indo Glob. J. Pharm. Sci., 2, 1–20, 2012.)*

10.8 Dermal products

Dermal products are used in many applications today and also serve as important assets in drug delivery. A rising concern for many individuals is cancer, specifically skin cancer. Since skin is the first level of protection, it is exposed to a lot of radiation and other insults, causing for there to be many diseases and conditions that need to be addressed. Dermal products are a very simple way to handle this issue. Building on the notion that nanotechnology is made for targeted treatment instead of holistic ones, these treatments are applied to the skin only where needed. One of the main drawbacks to using dermal products is that the penetration of the skin always proves to be troublesome. The Stratum Corneum is very hard to bypass and causes for lower quantity of the product and only superficial treatments. One of the ways to combat this is the use of vesicular systems. Transfersomes are examples of this; they have aqueous cores and are surrounded by complex lipid membranes. They are elastic and can thereby squeeze into irregular places and enter the skin [32]. Pharmacosomes are another novel vesicular lipid based micelle. They are used especially for insoluble botanical or herbal medicines [33]. The purpose of these novel drug delivery systems is to be able to administer a drug effectively, but also in a timely manner. This means that they have to be delivered at a specific rate defined by the needs of the body, and over the full period of the treatment [27]. Niosome is a class of molecular cluster formed by self-association of non-ionic surfactants in an aqueous phase. The unique structure of niosome presents an effective novel drug delivery system (NDDS) with ability of loading both hydrophilic and lipophilic drugs [34]. The final dermal-related nanosphere would be the ethosomes. Phospholipid based elastic nanovesicles that have a high content of ethanol (20–45%). They are made to be able to penetrate the skin, especially the Stratum Corneum layer which is the hardest to bypass. This is because the ethanol is an efficient permeation enhancer and can interact with polar heads of lipid molecules to increase the lipid fluidity and cell membrane permeability. They surpass the skin easier than other nanoparticles to deliver botanical products. However, dermal products are also much favored over other products if they can be effective since it a noninvasive procedure. It is quicker, cheaper, and has less side effects, such as the formation of scar tissue. With the rise of many nanotechnologies, such as ethosomes, the inefficiency of dermal administration is becoming less of a problem and making them more attractive to many industries [35]. Pharmacosome, liposome, transfersome, niosome, and ethosomes are all new forms of nanotechnologies, mostly in capsule form, that are applied dermally.

There are a variety of botanical drugs and herbal medicines that can be administered dermally by the novel vesicular drug delivery methods previously mentioned. These are drugs that can be packed in vesicles or other delivery modes,

Figure 10.6 *Product sample as dermal drug delivery. (From Poonam, V. and Pathak, K., J. Adv. Pharm. Technol. Res., 1, 274–282, 2010.)*

but are all delivered dermally. This noninvasive practice is very practical, easy, and cheap. Sertraline is a poorly soluble drug and has had troubles with oral delivery methods, so there have been new studies showing its success with transfersomes and dermal administration. Sertraline is an antidepressant drug that is a selective serotonin reuptake inhibitor. Transdermal delivery of it is a better-suited alternative. It avoids the extensive first-pass metabolism, gastro-intestinal disturbances such as nausea, dry mouth, diarrhea, and decreased appetite, among others. It also has poor bioavailability, requiring the drug be taken in high doses. When applied to the skin it is more convenient and safe and does not cause discomfort. These problems are very general and apply to most orally taken drugs, so having botanical drugs that can be administered dermally will be of great benefit [36] (Figure 10.6).

10.9 Cost effectiveness and other benefits

One of the major determining factors when it comes to advanced drug delivery methods versus conventional drug delivery methods is the cost effectiveness. This takes into consideration the cost and if it is worth it in comparison to it. Advanced drug delivery carriers such as nanocarriers and microcarriers can be difficult to manufacture and are costly to obtain. This is important to consider because of their size, structure, carriers capacity, biodegradability, and associated costs of research. However, these advanced drug delivery carriers have become extremely important and desirable features such as controlled release, target specific delivery, and drug stability [37]. Out of the many advanced drug delivery systems related to nanomaterials, polymeric nanoparticles are of particular due to their low cost [19].

Nanoparticles may reduce patient expenses when it comes to certain diseases. Among these diseases is cancer, a leading cause of death globally. Chemotherapy, radiation therapy, and biological therapies have been deemed unsatisfactory in terms of cost effectiveness. These methods are costly, have poor therapeutic efficacy, and are toxic to normal human tissues. These issues have ushered urgency in developing and advancing strategic drug delivery to improve the efficacy of treatment as well as reducing cost and side effects. The development of nanocarriers, nanoparticles, and nanotechnology have been designed to combat these issues as effective carriers for cancer therapeutic agents. Naturally occurring polymers have proven to be more advantageous than synthetic polymers in biomedical uses due to their versatile biocompatibility and low cost [17]. There are many other applications of nanotechnology aside from drug delivery, including forensic science, agriculture, medical therapeutics, space science, etc. [19].

The cost to benefit ratio of microparticle drug delivery has become more or less certain [39]. Through the advancement of nanotechnology, the materials necessary for its use has introduced increasing costs. However, the costs can be compensated through various applications. For nanoparticles with targeting ligands, coupling ligands with smaller weight and use of bio-responsive material directly to nanoparticles can allow for a smaller cost [39]. As nanotechnology becomes more optimal in clinical practice, the costs of its production and administration will be considered.

10.10 Conclusion

The use of herbal plants as therapeutic agents has taken a huge leap over the years. Extensive research and analyzation has allowed the invention of advanced drug delivery in effectively treating many medical complications. The use of combination of herbs has helped combat the problem of certain plant extracts not being effective due to low lipid solubility, inappropriate molecular size, low absorption, and poor bioavailability. An advanced drug delivery system having exceptional advantages, helps lower toxicity; increases pharmacological activity; enhances solubility, stability, and bioavailability; sustains delivery; improves tissue macrophages distribution; offers protection from chemical and physical degradation. Such drug delivery systems have a great potential in helping the advancement of science and plant-based drugs [38].

References

1. Jones, F.A. Herbs—Useful plants. 1996. *European Journal of Gastroenterology & Hepatology*, 8, 1227–1231.
2. Zhang, L. et al. Application of quality by design to The process development of botanical drug products: A case study. 2013. *AAPS Pharmscitech*, 14(1), 277.
3. U.S. Department of Health and Human Services Food and Drug Administration Center for Drug Evaluation and Research (CDER). What is a botanical drug?. 2017. *US Food and Drug Administration Home Page. https://www.fda.gov/downloads/Drugs/Guidances/UCM458484.pdf.*
4. Tiwari, G. et al. Drug delivery systems: An updated review. 2012. *International Journal of Pharmaceutical Investigation* 2(1), 2–11. *PMC*. Web.
5. Croy, S.R. et al. Polymeric micelles for drug delivery. 2006. *Current Pharmaceutical Design*, 12, 4669–4684.
6. Tomalia, D. et al. Starburst dendrimers: Molecular-Level control of size, shape, surface chemistry, topology, and flexibility from atoms to macroscopic matter. 1990. *Angewandte Chemie International Edition*, 29(2), 113–222.
7. Silvia, S. et al. Polymeric nanoparticles, nanospheres, and nanocapsules, for cutaneous applications. 2010. *Drug Target Insights*, 2, 147–157.
8. Pathak, Y.W. and Tran, H.T. Delivery of herbal extracts and compounds using nanotechnology. 2010. *Advances in Nanotechnology & Applications*, 2, 121–136.
9. Hans, M.L. and Lowman, A.M. Biodegradable nanoparticles for drug delivery and targeting. 2002. *Current Opinion in Solid State and Materials Science*, 6(4), 319–327.
10. Gastaldi, L. et al. Solid lipid nanoparticles as vehicles of drugs to the brain: Current state of the art. 2014 *European Journal of Pharmaceutics and Biopharmaceutics*, 87, 433–444.
11. Azubel, M. et al. Nanoparticle imaging. electron microscopy of gold nanoparticles at atomic resolution. 2014. *Science*, 345, 909–912.
12. Patela, J. et al. An overview of phytosomes as an advanced herbal drug delivery system. 2009. *Asian Journal of Pharmaceutical Sciences*, 4, 363–371.
13. Setty, A.R and Sigal, L.H. Herbal medications commonly used in the practice of rheumatology: Mechanisims of action, efficacy, and side effects. 2005. *Seminars in Arthritis and Rheumatism*, 34(6), 773–784.
14. Li, X.M. and Brown, L. Efficacy and mechanisms of actions of traditional chinese medicines for treating asthma and allergy. 2009. *Journal of Allergy and Clinical Immunology*, 123(2), 297–306.
15. Stickel, F. and Schuppan, D. Herbal medicine in the reatment of liver diseases. 2007. *Digestive and Liver Disease*, 39(4), 177–183.
16. Bennet, D. and Kim, S. Polymer nanoparticles for smart drug delivery. 2014. *Intech*. doi:10.5772/58422.
17. Li, Y. et al. Co-Delivery of drugs and genes using polymeric nanoparticles for synergistic cancer therapeutic effects. 2018. *Advanced Healthcare Materials*, 7(1), 1700886.
18. Torchilin, V.P. Recent advances with liposomes as pharmaceutical carriers. 2005. *Nature Reviews Drug Discovery*, 4, 145–160.
19. Kumari, A. et al. Biodegradable polymeric nanoparticles based drug delivery systems. 2010. *Colloids and Surfaces B: Biointerfaces*, 75(1), 1–18.
20. Soppimatha, K.S. et al. Biodegradable polymeric nanoparticles as drug delivery devices. 2001. *Journal of Controlled Release*, 70(1–2), 1–20.
21. Bin, L. et al. Bioactive SiO_2-CaO-P_2O_5 hollow nanospheres for drug delivery. 2016. *Journal of Non-Crystalline Solids*, 447, 98–103.
22. Guterres, S. et al. Polymeric nanopatricles, nanospheres and nanocapsules, for cutaneous applications. 2007. *Drug Target Insights*, 2, 147–157.
23. Chan, J. et al. Polymeric nanoparticles for drug delivery. 2010. *Cancer Nanotechnology. Methods in Molecular Biology (Methods and Protocols)*, 624, 163–175.

24. Zylberberg, C. and Sandro M. Pharmaceutical liposomal drug delivery: A review of new delivery systems and a look at the regulatory landscape. 2016. *Drug Delivery*, 23(9), 331–3329.

25. Meng C. et al. Liposomes containing cholesterol analogues of botanical origin as drug delivery systems to enhance the oral absorption of insulin. 2015. *International Journal of Pharmaceutics*, 489(1–2), 277–284.

26. Theresa, M. and Pieter R. Liposomal drug delivery systems: From concept to clinical applications. 2012. *Advanced Drug Delivery Reviews*, 65(1), 36–48.

27. Ajazuddin, S. Applications of novel drug delivery system for herbal formulations. 2010. *Fitoterapia*, 81(7), 680–689.

28. Kaushik, P. et al. Microencapsulation of omega-3 fatty acids: A review of microencapsulation methods. 2015. *Journal of Functional Foods*, 19, 868–881.

29. Singh, M.N. et al. Microencapsulation: A promising technique for controlled drug delivery. 2010. *Research in Pharmaceutical Sciences*, 5(2), 65–77.

30. Devi, V. et al. Importance of novel drug delivery systems in herbal medicines. 2010. *Pharmacognosy Reviews*, 4(7), 27–31.

31. Verma, H. et al. Herbal drug delivery system: A modern era perspective. 2013. *International Journal of Current Pharmaceutical Review and Research*, 4(3), 88–101.

32. Rai, S. et al. Transfersomes as versatile and flexible nano-Vesicular carriers in skin cancer therapy: The state of the art. 2017. *Nano Reviews & Experiments*, 8(1), 1–18.

33. Pandita, A. and Pooja S. Pharmacosomes: An emerging novel vesicular drug delivery system for poorly soluble synthetic and herbal drugs. 2013. *ISRN Pharmaceutics*, 10.

34. Moghassemi, S. and Afra H. Review: Nano-Niosomes as nanoscale drug delivery systems: An illustrated review. 2014. *Journal of Controlled Release*, 185, 22–36.

35. Poonam, V. and Pathak K. Therapeutic and cosmeceutical potential of ethosomes: An overview. 2010. *Journal of Advanced Pharmaceutical Technology & Research*, 1(3), 274–282.

36. Rajan, R. et al. Transfersomes—A vesicular transdermal delivery system for enhanced drug permeation. 2011. *Journal of Advanced Pharmaceutical Technology & Research*, 2(3) pp. 138–143.

37. Diego-Taboada, A. et al. Protein free microcapsules obtained from plant spores as a model for drug delivery: Ibuprofen encapsulation, release, and taste masking. 2012. *Royal Society of Chemistry*, 1(5), 707–713.

38. Bui, G.L.M, Le, U., Tran, H. and Pathak, Y.V. Delivery of herbal extracts and compounds using nanotechnology, Chapter 9. 2010. *Advances in Nanotechnology and Applications*, Pathak Y.V. and Tran H.T. (Eds.), 2, 121–136.

39. Cheng, Z. et al. Multifunctional nanoparticles: Cost versus benefit of adding targeting and imaging capabilities. 2012. *Science*, 338, 903–910.

40. Xu, W., Ling, P. and Zhang, T. Polymeric micelles, a promising drug delivery system to enhance bioavailability of poorly water-soluble drugs. 2013. *Journal of Drug Delivery*. Volume 2013, Article ID 340315, 15 pages. doi:10.1155/2013/340315.

41. Sharma, A. and Sharma, U.S. Liposomes in drug delivery: Progress and limitations. 1997. *International Journal of Pharmaceutics*, 154, 123–140.

42. Rudolph, A.S. The freeze-dried preservation of liposome encapsulated hemoglobin: A potential blood substitute. 2004. *Cryobiology*, 25, 277–284.

43. Anwekar, H., Patel, S. and Singhai, A.K. Liposome- as drug carriers. 2011. *International Journal of Pharmacy and Life Sciences*, 2, 945–951.

44. Nasongkla, N., Bey, E., Ren, J., Ai, H. et al. Multifunctional polymeric micelles as cancer-Targeted, MRI-Ultrasensitive drug delivery systems. 2006. *Nano Letters*, 6, 2427–2430.

45. Jones, M.C. and Leroux, J.C. Polymeric micelles—A new generation of colloidal drug carriers. 1999. *European Journal of Pharmaceutics and Biopharmaceutics*, 48, 101–111.
46. Cabral, H., Matsumoto, Y., Mizuno, K., Chen, Q. et al. Accumulation of sub-100 nm polymeric micelles in poorly permeable tumours depends on size. 2011. *A Nature Research Journal*, 6, 815–823.
47. Kim, B.S., Park, S.W. and Paula, T. Hammond. Hydrogen-Bonding layer-by-Layer-Assembled biodegradable polymeric micelles as drug delivery vehicles from surfaces. 2008. *ACS Nano*, 2, 386–392.
48. Jhaveri, A.M. and Torchilin, V.P. Multifunctional polymeric micelles for delivery of drugs and siRNA. 2014. *Frontiers in Pharmacology*, 5, 77.
49. Agnihotri, N., Mishra, R., Goda, C. and Arora, C.M. Microencapsulation—A novel approach in drug delivery: A review. 2012. *Indo Global Journal of Pharmaceutical Sciences*, 2, 1–20.

Intellectual Property Rights for Botanical Drug Products

Vishal Katariya and Kavita Gupta

Contents

11.1 Introduction to intellectual property (IP) rights

Botanical Drug Products (BDPs) are products of botanical origin often used as pharmaceuticals, cosmetics or nutraceuticals and diet supplements. According to the United States' Federal Food, Drug, and Cosmetic Act (FD&C), a botanical product can be a product marketed as one for diagnosing, mitigating, treating, or curing a disease; and which in general contains vegetable matter as ingredients. More importantly, those chemicals that may be obtained by purifying plant extracts via processes such as industrial fermentation to yield biopharmaceuticals do not fall under the category of botanical products. For every botanical product utilized as a dietary supplement or a cosmetic or as a pharmaceutical, the FDA has a set of separate regulations. An internationally used definition for a Botanical Drug Product is not available yet; however, most countries refer to such products as plant-based or herbal-based drug products that are further regulated based on their end use.

BDPs can be protected legally with the aid of Intellectual Property (IP) Rights. IPRs are a bundle of legal rights that are granted by the government of a country to an inventor or a creator for protecting the inventions or creations of the intellect. The basic principle of IP Rights is to provide an applicant with a set of exclusive legal rights that helps the inventor or creator to benefit and protect their work. Broadly, IP can be divided into two kinds—industrial property, which includes patents for inventions, trademarks, industrial designs and geographical indications; and copyright, which covers literary works (such as novels, poems and plays), films, music, artistic works (e.g., drawings, paintings, photographs and sculptures) and architectural design.[1] Intellectual property enables inventors and creators to not only protect their respective works but also gain monetary benefits from their products as well as earn recognition for them.

IP can be protected via any of the mentioned rights; however, the best mode for protecting BDPs would be to obtain a patent or keep it as trade secret. Most herbal or botanical products are derived from knowledge that has been passed down from generations, within communities and tribes, as indigenous remedies for malaises. These tend to be lesser known, and countries such as India and China, which have been using traditional medicines consisting of herbal or botanical concoctions since pre-historic times, have now developed libraries to record and store this traditional knowledge in order to protect it. Various international and national laws and committees, some of whom are discussed later in this chapter, protect biodiversity and traditional knowledge.

A patent, on the other hand, is an exclusive legal right granted for an invention (either a product or a process) that is novel in its conception and offers a new technical solution to any existing or pre-existing problems. Patents are most sought after because they provide an incentive to creators/inventors by recognizing their creativity in the form of a patent grant. Once a patent is granted the invention cannot be utilized, produced, distributed, or sold without the consent of the patent owner. A patent is a valuable form of legal protection for intangibles where protection is granted for a limited amount of time—in a majority of countries for a period of 20 years. Patents not only prove to be an incentive for creators or inventors, by providing the possibility of monetary benefits for their marketable inventions, but also tend to encourage innovation and help improve the quality of life of individuals.

In order to obtain a patent for an invention one must begin by filing an application for a patent. The application is written in a clear and precise manner with enough detail to enable a person, with average intelligence and a basic understanding of the field, to use or reproduce the said invention. A patent application generally contains the title of the invention followed by a short description of the field of invention. Most countries have varying structures for their patent applications but all of them commonly contain a description of the background and of the invention. Such descriptions can include drawings, plans or diagrams to illustrate the content in greater detail. An important section of the application is the "claims" where the inventor or creator claims certain specific features of their invention, which details the extent of protection to be granted by the patent.

Patent owners have the advantage of deciding who can or cannot use their patented inventions; however, this is only during the term of the patent duration. Patent owners can also grant licenses to other parties for using, manufacturing, and distributing their inventions or sell their rights to the inventions to another party. During the patent term, the patent is made available to the public as a publication in the Official Gazette of a country, however once it expires the patent is no longer protected and the invention enters the public domain where the owner has no exclusive rights to the invention anymore.

Ideally, an invention must, in general, fulfill the following conditions in order to be protected by a patent: It must have practical use; it must be novel, i.e., it should contain some new element or characteristic that does not form a part of the existing knowledge (called prior art) in its particular technical field; consist of an inventive step that cannot be presumed by a person with average knowledge of the technical field; and lastly the subject matter must be accepted as "patentable" under law.[1] Patenting comes with many limitations, and the concept "everything under the sun can be patented" does not hold true. There are many countries that have an expansive list of what cannot be patented, which differ from each country. Some of the things that are restricted are: a scientific theory, any mathematical methods, various natural plant or animal varieties, discovering a natural substance, any method for commercial business or medical treatment.

There are various routes to obtain a patent, such as a national application, a regional application, and via the Patent Cooperation Treaty (PCT) regime. Ideally, the concept of an international patent does not exist. One could obtain a patent at any one or more national office by applying to each country they would want to obtain a patent in or one could chose to send in an application to a regional office such as the European Patent Office (EPO) or the African Regional Intellectual Property Organization (ARIPO), which eliminates the need to go to each country individually and file a patent. Once a creator/inventor decides to patent an invention and prepares a patent application, he/she can file the application at any national patent office or a regional patent office. Regional patent office services provide the advantage of requesting for a patent protection in one or more countries (that are a member of the regional office) and each country has the discretion to grant or reject a patent within its borders. The drawback however is that regional offices provide protection only in countries that fall within their regional offices. On the other hand the WIPO-administered Patent Cooperation Treaty (PCT) allows the applicant to file a single patent application that has the same effect as national applications filed in the designated member countries. The advantage of filing a PCT application is that an applicant seeking protection may file one application and request protection in as many signatory states as required.

Another effective strategy to protect a botanical product from competition involves a type of IPR called trade secret. A trade secret essentially consists of vital information related to the product (or process), which is not disclosed to the public and is maintained as a secret for as long as possible and desired. Technically any kind of confidential information that provides a company with a competitive edge is considered a trade secret, and they can consist of manufacturing or industrial secrets as well as commercial secrets.[2] The legal protection of trade secrets usually falls under the general concept of protection against unfair competition or the protection of confidential information. The subject matter of trade secrets is usually very vast and can include methods for sales and distribution, consumer profiles, advertising strategies, lists of suppliers and clients, and manufacturing processes. It is not always easy to distinguish what specifically falls under the scope of trade secrets, but the content is always circumstantial and judged individually. Unlike other IPRs, trade secrets have limited protection because a trade-secret holder is only protected from unauthorized disclosure or use, often referred to as misappropriation. Any unfair practice with respect to trade secrets usually includes industrial or commercial espionage, breach of contract and/or breach of confidence. However, if maintained, trade secrets do not expire unlike patents, which have a set term of protection (usually 20 years) and can remain indefinitely until discovered or lost. There is no legal framework in most countries for trade secrets however, in the US every state has the liberty to form its own set of rules[3] and currently several US states have adopted the Uniform Trade Secret Acts (UTSA). According to the World Intellectual Property Organisation (WIPO), one of the 17 specialized agencies of the United Nations responsible for promoting the protection of intellectual property throughout the world through cooperation among member countries of the organization; the unauthorized use of such information by persons other than the holder is regarded as an unfair practice and a violation of the trade secret.

For BDPs in particular, there are certain guidelines and regulations laid out by each country for granting a patent as well as for marketing rights. These will be explained in further sections. It is important to note that BDP's do not have a firm set of rules or legislations set in place and are still in the developmental stages around the world on the national and international fronts.

11.2 Market exclusivity

In the case of Botanical Drug Products the laws of market exclusivity would be applicable only in case it can be classified as a pharmaceutical drug. Pharmaceutical development is an expensive, time-consuming and uncertain process that takes years to complete. A pharmaceutical drug will usually have two kinds of legal protections, namely exclusivity and patent protection. Exclusivity is a creation of law and enables the drug product to have exclusive or monopoly status in the market for a certain number of years. Patent is a government authority or license conferring a right or title for a set period, especially the sole right to exclude others from making, using, or selling an invention.

Often, patent protection expires before a new drug is approved for marketing. As a result, most pharmaceutical companies depend on the exclusivity rights granted by the appropriate authorities to recoup their considerable investment in the drug development and approval process in order to succeed in the global marketplace. Pharmaceutical companies generally obtain patents on their products or processes long before their product candidates are ready to go to market as it can take up to 12 years for a company to obtain market approval and there is often little, if any, patent protection left on the product at the time of marketing. To provide pharmaceutical companies with an opportunity to recoup their investment in drug research and development, and to incentivize continuing innovation, many countries have implemented numerous provisions to extend the period during which companies can market their drugs free of generic competition. These non-patent exclusivity provisions allow pharmaceutical companies to market products without competition from incoming generics, resulting in significant financial benefits for the original drug manufacturer. It is essential that a pharmaceutical company evaluate its exclusivity options and develop its competitive strategy early in the drug development process. The period of exclusivity differs from country to country.

In the United States, the act governing exclusivity laws is the US Federal Food, Drug and Cosmetic Act (FD&C) and provides that the FDA cannot legally approve an application of a generic drug for that product until the exclusivity period expires. Drugs in general, can be marketed in the US in the following forms—New Drug Application (NDA) and an Abbreviated New Drug Application (ANDA). An NDA application contains "full reports" relating to safety and efficacy investigations whereas an ANDA is submitted to acquire marketing rights for a duplicate, or generic, version of an already approved drug where no safety and efficacy data needs to be submitted for approval but data pertaining to bioequivalence is a must. NDA also contains information regarding any drug products as well as their methods of use. These patents are further enlisted in the Approved Drug products with Therapeutic Equivalence Evaluations, also known as the Orange Book. This period of market exclusivity is independent of any patent protection that might be available upon grant of a patent and is granted to an NDA applicant as follows:

- Orphan Drug Entity (ODE) (a product for treating fewer than 200,000 patients in the US per year)—7 years
- New Chemical Entity (NCE)—5 years
- "Other" Exclusivity—3 years for a product other than a NCE (dependant on certain conditions and criteria)
- Pediatric Exclusivity (PED)—6 months added to existing Patents/Exclusivity
- Patent Challenge (PC)—180 days (this exclusivity is for ANDAs only)

In Europe, the European Medicines Agency (EMA) oversees the regulations pertaining to botanical drug products marketed as pharmaceuticals. In the past few years, the EU has expanded significantly the opportunities for drug manufacturers to obtain market exclusivity for their products, and since 1993, European drug companies have been able to obtain a supplementary protection certificate (SPC) to extend for up to 5 years the patent for certain medicinal products marketed in the EU in order to compensate them for the lengthy time period required to obtain regulatory approval of these products.

An SPC will be granted only if, at the date of application, the product

1. Is protected by patent.
2. Is the subject of the first valid marketing authorization granted to market the product for a medicinal use.
3. Has not already been the subject of an SPC.

The SPC comes into force only after the corresponding general patent expires for a period equal to the period that elapsed between the date on which the patent application was filed and the date of the first marketing authorization, minus 5 years. The SPC term may not exceed 5 years from the date on which it takes effect. In 2005, the EU Data Exclusivity Directive 31 was brought into force under which, sponsors may receive up to 11 years of exclusivity for new drugs. The exclusivity available under the Directive may include 8 years of data exclusivity, two years of marketing exclusivity, and a potential one-year extension. In addition, under the EU Orphan drug regulations, which became effective in 2000, the Community and the Member States may not accept or grant for 10 years, a new marketing authorization, or an application to extend an existing marketing authorization, for the same therapeutic indication as an orphan drug. Finally, Paediatric Regulation, which became effective in 2007, provides sponsors with the right to apply for a six-month extension to the product's SPC in return for conducting paediatric studies on the product.

In Australia, market exclusivity laws are governed by the Australian orphan drugs policy, which was set up in 1997. The main characteristic of the Australian Orphan Drugs Program is based upon a close collaboration of the TGA with the US FDA and takes into account the FDA's orphan drugs evaluations and aims at providing a 5-year exclusivity period to orphan drugs but this is still under consideration by the Australian jurisdiction. However, the concept of market exclusivity is relatively new for the Australian markets and is still under development as substantial laws governing this concept are still not available.

In India, the concept of data exclusivity prevails and not market exclusivity. Data exclusivity is a form of exclusivity which prevents regulators from using the clinical trial test data that had been used to approve the originator's product, to also approve the chemically (or otherwise) equivalent generic product. This means that if the generic company wants to get regulatory approval during the period of data exclusivity (generally 5–10 years), it needs to duplicate the expenses and time taken to take its product through clinical trials. If the period of data exclusivity completely overlaps with the patent duration, there is usually no substantial effect as the patent would anyway prevent generics from releasing the product. However, off-patent or non-patented products can also be granted data exclusivity—in which case they would enjoy 5–10 years of exclusivity. Data-exclusivity rights are also not necessarily bound by the same exceptions and flexibilities that patent rights are.

11.3 Country specific regulations for IPR and BDPs

BDP's are increasingly becoming popular in the intellectual property world as awareness relating to botanical products marketed as a multitude of drug varieties has spread across the globe. Many of the botanical products used today have their sources dating back to eras when such use was termed as traditional knowledge. Herbal medicines have been used in India and China since time immemorial, and so have the tribes of Africa and the Middle East. In order to spread awareness regarding techniques and products used by the locals of these countries, certain legal provisions have been set in place to protect this traditional knowledge. An example would be the Kingdom of Thailand, which has an act for the "Protection and Promotion of Traditional Thai Medicinal Intelligence" along with a committee for the same that protects recipes and texts drawn from traditional Thai medicine. Safeguarding traditional knowledge is important for countries with a rich background relating to the use of herbal and botanical products since before the concept of IP rights. Enacting stringent laws is another strategy an example of which would be the South American country of Peru, which has legislated to protect indigenous people's collective knowledge but also permits knowledge licensing contracts. The country's Cusco region has outlawed exploitation of native species for commercial gain, including patenting genes or other resources.

The Convention on Biological Diversity (CBD), an international convention signed by 150 government leaders at the 1992 Rio Earth Summit, understands and recognizes the importance of traditional knowledge for enhancing the conservation and sustainable use of biodiversity. Article 8 (j) of the convention relates to Traditional Knowledge, Innovations and Practices which explicitly states that -

> *"Each contracting Party shall, as far as possible and as appropriate: Subject to national legislation, respect, preserve and maintain knowledge, innovations and practices of indigenous and local communities embodying traditional lifestyles relevant for the conservation and sustainable use of biological diversity and promote their wider application with the approval and involvement of the holders of such knowledge, innovations and practices and encourage the equitable sharing of the benefits arising from the utilization of such knowledge innovations and practices."*[4]

Patent cases in relation to the neem tree, turmeric, quinoa, and Ayahuasca are prominent examples of how IP rights were granted to a few research companies' in spite of predominant knowledge about the use of those plants within indigenous communities. The novelty and benefit aspects for such patents now follow strict regulations and guidelines set up by many countries around the world, some of which are explained below.

11.3.1 United States of America

Botanical products are classified as a drug, food or a dietary supplement by the United States Food and Drug Administration (FDA) on the basis of the claims or end use. An article is generally a food if it is used for food according to 21 U.S.C. 312(f)(1) and can be used as a drug, medical device, or cosmetic depending on its "intended use."[5] The intended use of a product is created by claims made by the manufacturer/distributor of the botanical product usually via advertisements or labels or even orally.

A botanical product that is used to prevent, diagnose, mitigate, treat, or cure a disease would fall under the category of a drug whereas if the intended use of the botanical product is to affect the structure or the function of the human body, it can be classified as a drug or a dietary supplement. As per the FDA, a drug must be marketed under an approved New Drug Application (NDA). Botanical drug products, which are marketed as dietary supplements, on the other hand, are regulated in the US by the Dietary Supplement Health and Education Act of 1994 (DSHEA), which is a 1994 statute of United States federal legislation that defines as well as regulates dietary supplements. Under this particular act, dietary supplements are effectively regulated by the FDA for Good Manufacturing Practices (GMP) under 21 CFR Part 111. Off late there has been a tremendous increase in the US for patents of herbal or botanical drug products that are derived from traditional preparations. According to statistics, between the years 1982 and 2007, more than 350 botanical investigational new drug (IND) applications and pre-IND consultation requests were submitted to the agency.[6]

The United States recognizes botanical products with health-related claims as those products that can be marketed as conventional foods, dietary supplements, or drugs, depending on the specific claim, as described in the Dietary Supplement Health and Education Act (DSHEA) of 1994.[7] Conventional food and dietary supplements that do not claim any use for treatment of diseases are usually regulated by the Centre for Food Safety and Applied Nutrition of the FDA. In the US, the FD&C defines a drug as "*an article intended to mitigate, treat, cure, diagnose or prevent a disease or its related symptoms (disease claim), or as an article intended to affect the structure or function of the body (structure-function claim).*"[8] On the other hand, a dietary supplement is considered a drug only if it bears a disease claim according to DSHEA. For such situations the dietary supplements are regulated by the FDA's Centre for Drug Evaluation and Research (CDER).

The FDA regulates botanical as well as non-botanical drugs the same way. The marketing of botanical drugs can be done if and only if they meet the legal requirements for demonstrating the safety and effectiveness of a new drug in accordance with the relevant sections of the FD&C. They must also satisfy the manufacturing requirements in order to ensure high product quality. Those botanical products which are marketed without disease claims under DSHEA do not need to submit data relating to clinical tests to the FDA. The difference between disease claims and structure-function claims (often referred as "health claims") is defined clearly in the FDA Final Rules of January 5, 2000.[9]

The FDA has also released a draft on industry guidelines for botanical drug products,[10] available online, which explains explicitly the regulatory approaches as well as the marketing strategies for botanical products as NDAs. This industry guideline does not address those drugs that contain animals or animal parts and/or minerals (alone or in combination) with botanicals; however, it describes that when a drug product contains botanical ingredients in combination with either a synthetic or highly purified drug OR a biotechnology derived or other naturally derived drug, the industry guidance would then apply to only the botanical portion of the product.

11.3.2 India

In India, herbal drugs are regulated under the Drug and Cosmetics Act (DCA) 1940 and Rules 1945, where the regulatory provisions for Ayurveda, Siddha, and Unanni medicines are provided vividly. The department of AYUSH (Department of Ayurveda, Yoga and Naturopathy, Unani, Siddha and Homoeopathy)[11] is a regulatory governmental authority created in March 1995 as the Department of Indian Systems of Medicine and Homoeopathy (ISM&H), which mandates that any manufacture or marketing of herbal drugs has to be done only after obtaining manufacturing licenses as applicable.[12] The official Indian pharmaco-poeias and formularies provide a better understanding of the standard of quality of the medicines marketed in India.

The D and C Act extends the control over the licensing, formulation composition, manufacture, labelling, packing, quality and export of products. An important section of the Act is Schedule T, which lays down the Good Manufacturing Practice (GMP) standards to be followed when manufacturing herbal medi-cines. The first schedule of the D and C Act has listed authorized texts which should be followed for licensing any herbal product under the two categories—ASU drugs, and patent or proprietary medicines.

India's Council for Scientific and Industrial Research (CSIR), in 2001, launched a Traditional Knowledge Digital Library (TKDL) which contains a list of indig-enous remedies and medicinal plants. CSIR has signed agreements with leading international patent offices, such as the EPO, United Kingdom Trademark & Patent Office (UKPTO), and the United States Patent and Trademark Office to protect traditional knowledge from biopiracy, by enabling free access to pat-ent examiners at international patent offices to the TKDL database for patent search and examination.[13]

11.3.3 China

Traditional Chinese medicines (TCM) have been used for centuries for treat-ments in China, and on April 16, 2012 the Good Practice in Traditional Chinese Medicine (GP-TCM) Research Association was officially launched at the GP-TCM Congress in Leiden, the Netherlands.[14] GP-TCM is the EU's first coordination action dedicated to TCM research. The Association is funded by the European Commission under its 7th Framework Programme (FP7), and the project has engaged more than 200 scientists and clinicians from 112 institutions in 24 countries in discussions on good practice issues related to various aspects of Chinese herbal medicine and acupuncture research.

The GP-TCM has listed 10 major objectives[15] on their website, which are as follows:

1. *Perpetuate the interactive network established by the FP7 GP-TCM consortium.*
2. *Promote discussion and implementation of good practice in TCM research and development, including the use of sustainably sourced materials.*
3. *Advocate high-quality evidence-based research and develop-ment on TCM as well as on its integration with conventional medicine.*
4. *Organize and co-organize scientific meetings.*

5. *Nurture young TCM researchers at different levels in an interdisciplinary approach, including BSc, MSc, PhD and post-doctoral programmes.*
6. *Facilitate collaborations and sharing of resources, expertise and good practice among members, industry and regulatory agencies.*
7. *Encourage collaborations with existing relevant societies, consortia and organizations.*
8. *Strengthen interdisciplinary, interregional, and intersectoral collaborations in TCM research and development.*
9. *Perpetuate good practice in publishing TCM research outcomes.*
10. *Disseminate scientific research outcomes and latest developments in regulatory sciences to stakeholders, industry, professional groups and the public.*

Traditional herbal medicines are regulated in China and Taiwan by a different set of regimes in comparison to those that are followed for chemical and biological drugs. The manufacturer of an herbal product can also claim for efficacy to treat a condition if it is based on historical literature provided in the Chinese Classics Eastern Literature. The product must also meet the good manufacturing requirements as defined by the regulatory authorities for TCM. As is observed in all countries with relation to drugs, the Chinese State Food and Drug Administration (SFDA) and the Taiwanese Department of Health (DOH) have implemented regulations on Good Manufacturing Practice (GMP) guidelines similar to those for pharmaceutical products for manufacturers of herbal medicines to follow. Generally a manufacturer is not required to provide data relating to the safety of the herbal product because the long withstanding history of human exposure to the herbs has provided enough proof of stability and safety for human use.

Needless to say, however, these guidelines on botanical or herbal drug products need further dissemination and development through continued inter regional as well as international interdisciplinary collaborations such as the GP-TCM.

11.3.4 Japan

Traditional Japanese medicines have been in use for many years and are still practiced and used today. Conventionally traditional Japanese medicines can be divided into folk medicines and Chinese medicines (or Kampo as is predominantly called in Japan). Kampo drug products are registered as drugs by the Ministry of Health and Welfare (MHW) in Japan. The acceptance of Kampo drugs used to occur without the need for submitting any clinical validation studies. The new Kampo drug products, however, are regulated and monitored in essentially the same way as the Western drugs are in Japan. The data required to be submitted for new Western drugs are required for new Kampo drugs as well, which include data from three-phase clinical trials as well.

11.3.5 Malaysia

In Malaysia, products of herbal origin fall under the category of regulated products and can be marketed as either traditional products or health supplements. Any individual who markets herbal products and intends to place them for sale in the market is mandatorily required to register their product with the Malaysia Registrar of Business or Suruhanjaya Syarikat Malaysia (which came into operation on April 16, 2002). There is a mandate of labeling the phrase "traditionally used for" only in front of any claim that is made on traditional products. Functional claims, however, are permitted for supplements.

11.3.6 Philippines

Botanical products utilized as herbal medicines are marketed in the Philippines as traditionally used herbal products. The Bureau of Food and Drugs (BFAD)[16] is the regulatory body in the country and mandates that the traditionally used herbal products must be registered before manufacturing, importing, or marketing the product. The BFAD regulation also looks over the brand names of the traditional herbal products as prior clearance by the body is required before a person can register the product. The regulatory bodies of the Philippines have a list of requirements to be fulfilled for marketing botanical products as traditionally used herbal products; some of these are related to the preparation of the product and state that the preparation should be from plant materials and the claimed application of the product should be based only on traditional experience of long usage i.e., it should be at least five or more decades as documented in medical, historical, and ethnological literature.[17]

Furthermore, the authentication of the plant specimen, used in the product, needs to be obtained from the Philippine National Museum or any BFAD-recognized taxonomist. Similarly for imported products the certificate of authenticity of the plant must be obtained from the authorized government agency of the country of origin. National regulatory standards for manufacturing are to be followed and the product indications should not be administered by a physician.

11.3.7 Nigeria

In Nigeria the trade of herbal products is regulated by the National Agency for Food and Drug Administration and Control (NAFDAC). Herbal products are classified by the NAFDAC as Herbal Medicines and Related Products. It is mandatory to register herbal medicines and related products prior to marketing them in Nigeria,[18] and all advertisements require a preclearance from the NAFDAC. Furthermore, no advertisements can be made for the product as a cure for any disease conditions enlisted in Schedule 1 of the Food and Drug Act 1990.

11.3.8 Saudi Arabia

The document titled "Regulations for Registration of Herbal Preparations, Health and Supplementary Food, Cosmetics and Antiseptics that have Medicinal Claims" is issued by the Ministry of Health of the Kingdom of Saudi Arabia,[19] and it mandates that the formal application for registration, submitted to the General Directorate of Medicinal and Pharmaceutical Licences at the Ministry of Health, should be based upon the registration of the product in the country of origin. Products of herbal content or origin can be registered as traditional products and are permitted in the country if and only if they have at least 50 consecutive years of traditional use. The dosages and method of preparation for such products must relate to those traditionally existing. Moreover, further classification of products based on the evidence provided can be as either Pharmacopoeial evidence for traditional products or Nonpharmacopoeial evidence for traditional products.

For Pharmacopoeial evidence produced for traditional products, the medicinal ingredients, quantity, recommended dose, route of administration, duration of use, dosage form, directions of use, risk information should be same as the Pharmacopoeia whereas the method of preparation should be traditional. For the latter category, any two independent references must be provided to supplement the evidence to support the safety and efficacy of the product—from clinical studies, pharmacopoeias, textbooks references, peer-reviewed published articles, data from nonclinical studies on pharmacokinetics, pharmacodynamics, toxicity information, reproductive effects, and the potential genotoxicity or carcinogenicity of an ingredient or information based on previous marketing experience of a finished product.

11.3.9 Australia

Botanical drug products are regulated in Australia by the Therapeutic Goods Administration, which is the regulatory agency of Australia responsible for regulating herbal products under the category of complementary medicine. Alternative types of medicines such as Ayurvedic medicine, traditional Chinese medicine as well as Australian indigenous medicines also fall under this category. The Therapeutic Goods Administration, in general, handles the regulation for therapeutic goods including prescription medicines, vaccines, sunscreens, vitamins and minerals, medical devices, and blood and blood products. Usually, complementary medicines that do not require medical supervision are permitted but must be registered on the Australian Register for Therapeutic Goods (ARTG)[20] before marketing. Low-risk medicines are mandatorily required to be listed whereas those medicines used for comparatively higher risk conditions must be registered on the ARTG. An important condition is that only evidence-based claims that are entered on the ARTG will be permitted.

11.3.10 Canada

Health Canada, a part of the Federal Department, has been regulating herbal and traditional medicines since January 1, 2004 under the Natural Health Products (NHP) regulations.[21] The NHP regulations mandate that a manufacturer, packer, labeller, or importer needs to have a prior registration with Health Canada before commencing any related activity. The process involves the registration of the manufacturing site/s and the products as well as submitting various documents to the Natural Health Product Directorate (NHPD) relating to the product composition, standardization, stability, microbial, and chemical contaminant testing methods and tolerance limits, safety, and efficacy along with ingredient characterization, quantification by assay or by input. The authority also mandates that NHPs must comply with the contaminant limits, and they must be manufactured in accordance with the GMP norms.

11.3.11 European Union

The regulatory body in the EU, which is the European Medicines Agency,[22] has laid down the following two ways to register herbal medicinal products:

A full marketing authorization by submission of a dossier, which provides the information on quality, safety, and efficacy of the medicinal products including the physicochemical, biological, or microbial tests and pharmacological, toxicological and clinical trials data; under directive 2001/83/EC;

For traditional herbal medicinal products, which do not require medical supervision, where evidence of long traditional of use of medicinal products exists and adequate scientific literature to demonstrate a well-established medicinal use cannot be provided; a simplified procedure under directive 2004/24/EC exists.

For the EU, like most other countries, the evidence of traditional use is an acceptable form of evidence to prove the efficacy of the product; however, it is plausible that authorities may ask for relevant data to support the safety of the product. The quality control guidelines mandate that the physicochemical and microbiological tests must be provided with product specifications. The product should also comply with the quality standards in relevant pharmacopoeias of the member state or the European Pharmacopoeia. Also, the bibliographic evidence provided should support that the product has been in medicinal use for at least 30 years (outside the EU and includes usage for at least 15 years within the EU). The application for traditional use registration is handed over to the Committee for Herbal Medicinal Products (HMPC)[23] in case the product has been in the community for less than 15 years however; it also qualifies for the simplified registration procedure under the directive.

11.4 International IP regulations for BDP
11.4.1 The Nagoya Protocol

The Nagoya Protocol on Access to Genetic Resources and the Fair and Equitable Sharing of Benefits Arising from their Utilization to the Convention on Biological Diversity[24] is an international agreement that aims at sharing the benefits arising from the utilization of genetic resources in a fair and equitable way. The Nagoya Protocol on Access to Genetic Resources and the Fair and Equitable Sharing of Benefits Arising from their Utilization was adopted by the Conference of the Parties to the Convention on Biological Diversity at its tenth meeting on October 29, 2010 in Nagoya, Japan. In accordance with its Article 32, the Protocol was opened for signature from February 2, 2011 to February 1, 2012 at the United Nations Headquarters in New York by Parties to the Convention. So far there are 68 countries party to the protocol with 70 ratifications and over 92 signatures. The most prominent parties to the protocol are India, European Union, South Africa, and UAE.

The protocol came in to force on October 12, 2014 as a supplement to the Convention on Biological Diversity (CBD) and its main objective is to equitably share the benefits gained from using genetic resources as well as provide the associated rights of indigenous communities. This protocol is an international attempt to ensure that anyone under the jurisdiction who benefits from traditional knowledge has obtained prior informed consent from the respective countries and have negotiated a fair as well as equitable deal to share those benefits.

One of the most prominent protocols is Article 10, which focuses on benefit sharing. Article 10 of the Protocol states that:

> *Parties shall consider the need for and modalities of a global multilateral benefit-sharing mechanism to address the fair and equitable sharing of benefits derived from the utilization of genetic resources and traditional knowledge associated with genetic resources that occur in transboundary situations or for which it is not possible to grant or obtain prior informed consent. The benefits shared by users of genetic resources and traditional knowledge associated with genetic resources through this mechanism shall be used to support the conservation of biological diversity and the sustainable use of its components globally.[25]*

A simple example is the Kani tribe of Southern India where an agreement following the protocol's guidelines ensures that the tribe receives a just share of income arising from research by the Tropical Botanical Garden and Research Institute

(TBGRI) on the Arogyapacha plant (*Trichopus zeylanicus travancoricus*) which is traditionally used to revitalize and is now licensed and sold as "Jeevani."

11.4.2 WIPO: The Intergovernmental Committee on Intellectual Property and Genetic Resources, Traditional Knowledge and Folklore

In the fall of 2000 the WIPO established the WIPO Intergovernmental Committee on Intellectual Property and Genetic Resources, Traditional Knowledge and Folklore (hereinafter "Intergovernmental Committee" or IGC).[26] Essentially the mandate of the IGC is to facilitate the discussion of intellectual property issues that arise in the context of:

1. Access to genetic resources and benefit sharing
2. Protection of traditional knowledge, innovations and creativity
3. Protection of expressions of folklore, including handicrafts[27]

The IGC was established in September 2000 and serves as an international forum where WIPO member states can discuss the intellectual property issues that arise in the context of access to genetic resources and benefit sharing as well as the protection of traditional knowledge and traditional cultural expressions. In 2009 the WIPO members decided that the IGC should establish an agreement, by starting formal negotiations, on one or more international legal instruments that would ensure the effective protection of genetic resources, traditional knowledge and traditional cultural expressions. Such an agreement could either be established as a recommendation for all the WIPO members or become a formal treaty that would bind the member countries which would choose to ratify it.

Since its inception the IGC has stimulated an increased recognition of traditional knowledge (TK) within the patent system, and in 2002 certain TK journals were included in the minimum documentation for applications under WIPO's PCT. Also, TK classification tools were integrated within the International Patent Classification in the year 2003. Further progress was made in 2002 when the IGC accepted technical standards for the documentation of TK, which was developed at a WIPO meeting in Cochin, India.

In order to provide guidance on the IP aspects of mutually agreed terms for fair and equitable benefit-sharing related to Genetic Resources (GR), WIPO has developed, and regularly updates, an online database of relevant contractual practices. A rough draft consisting of guidelines on IP clauses relating to access and benefit-sharing agreements have been developed. Under the IGC, WIPO has also carried out numerous studies and developed many other resources (such as glossaries, surveys of national experiences, a laws database and training programmes), which have proved useful for member states and others. Negotiations on TK have continued to this date, and the WIPO member states may in due course decide to convene a diplomatic conference for the final adoption of one or more international instruments.

11.4.3 International Union for the Protection of New Varieties of Plants

Botanical products do not follow a strict regulatory procedure, but one of the factors that affects IPRs relating to BDPs are plant variety regulations. Internationally, approximately 72 countries, including USA, EU, China, and Central Africa, are party to the International Union for the Protection of New Varieties of Plants (UPOV), which is an intergovernmental organization with its headquarters in Geneva, Switzerland. India has the Protection of Plant Varieties and Farmers' Rights (PPVFR) Authority which was established as an effective system for protecting plant varieties, the rights of farmers as well as plant breeders.

UPOV was established by the International Convention for the Protection of New Varieties of Plants and was adopted in Paris in the year 1961. UPOV's mission is to provide and encourage an efficient system of protection for plant varieties. The aim of the UPOV is to encourage the development of new varieties of plants and provide the basis for member countries to encourage plant breeding by granting breeders of new plant varieties an intellectual property right called the breeder's right.

Breeder's right is important because it means that the authorization of the breeder is required to propagate the variety for commercial purposes. The UPOV Convention describes and defines the acts that require the breeder's authorization when propagating material of a protected variety and, under certain conditions, for the harvested material. UPOV members may also decide to extend the protection to products which are made directly from the harvested material under certain conditions. In order to obtain this protection, the breeder must file an individual application with the authorities of the respective UPOV members entrusted with the task of granting the breeders' rights. Under the UPOV Convention, the breeder's right is granted only when the variety is new; distinct; uniform; stable and has a suitable denomination.

In India, the PPVFR encourages the development of new varieties of plants that may be considered as essential to recognize and protect the rights of the farmers with respect to their contribution made at any time in conserving, improving, and making available plant genetic resources for the development of the new plant varieties.[28] The PPVFR Authority is of the view that in order to accelerate agricultural development it is necessary to protect the plants breeder's rights as well so as to encourage investment for research and development purposes of new plant varieties. India, having ratified the Agreement on Trade Related Aspects of the Intellectual Property Rights (TRIPS), had to make a provision for giving effect to the Agreement. To give effect to the aforesaid objectives, the Protection of Plant Varieties and Farmers' Rights (PPVFR) Act, 2001 was enacted in India.

11.5 Conclusion

The regulatory bodies in every country across the globe monitor and regulate BDP's on the basis of their end usage and their marketing strategies. Any product of botanical content that is marketed as a pharmaceutical product is subjected to the market exclusivity protection available for pharmaceutical drugs in most countries. Along with the constantly evolving status of BDPs in the scientific arena the laws relating to such herbal products are concurrently developing. Today BDPs can be marketed and protected across the globe under various legislations depending upon the nature of the product and the class under which the manufacturer or retailer wishes to market it. Most BDPs are sold as dietary supplements and cosmetics (over the counter products). However, this does not prevent them from being marketed as pharmaceutical products thus, providing the necessary protection under pharmaceutical laws when compared to those in place for cosmetic or nutraceutical products. Thus, the manufacturers of BDPs have a plethora of opportunities to explore various markets, with adequate protection being provided by the laws across most developed and developing countries today.

Endnotes

1. http://www.wipo.int/edocs/pubdocs/en/intproperty/450/wipo_pub_450.pdf
2. http://www.wipo.int/sme/en/ip_business/trade_secrets/trade_secrets.htm
3. Kewanee Oil Co. v. Bicron Corp., 416 U.S. 470, 94 S.Ct. 1879, 40 L.Ed.2d 315 (1974)
4. https://www.cbd.int/traditional/
5. 21 U.S.C. 312(g)(1)(B) and (C), (h)(2) and (3), (i)
6. Chen T Shaw, Dou J et al. New therapies from old medicines. Nature Biotechnology 26, 1077–1083; 2008
7. https://ods.od.nih.gov/About/DSHEA_Wording.aspx#sec3
8. See Sub-chapter II, S.321—(g)(i) : https://www.gpo.gov/fdsys/pkg/USCODE-2010-title21/pdf/USCODE-2010-title21-chap9-subchapII-sec321.pdf
9. http://www.fda.gov/Food/GuidanceRegulation/GuidanceDocumentsRegulatory Information/DietarySupplements/ucm103340.htm
10. http://www.fda.gov/downloads/drugs/guidancecomplianceregulatoryinformation/guidances/ucm070491.pdf
11. http://indianmedicine.nic.in/index.asp?lang=1
12. Malik V, (ed.); Law Relating to Drugs and Cosmetics (23rd ed. Lucknow, UP: Eastern Book Company; 2013)
13. http://www.tkdl.res.in/tkdl/langdefault/common/Abouttkdl.asp?GL=Eng
14. http://www.gp-tcm.org/
15. http://www.gp-tcm.org/about/objectives/
16. http://www.doh.gov.ph/node/932
17. http://apps.who.int/medicinedocs/pdf/whozip57e/whozip57e.pdf
18. http://www.nafdac.gov.ng/attachments/article/205/13_Herbal_Medicines_And_Related_Products__Registration__Regulations_2004.pdf
19. http://old.sfda.gov.sa/en/drug/drug_reg/Pages/drug_reg.aspx

20 https://www.tga.gov.au/tga-basics
21 http://www.hc-sc.gc.ca/dhp-mps/prodnatur/legislation/acts-lois/prodnatur/index-eng.php
22 http://www.ema.europa.eu/ema/index.jsp?curl=pages/regulation/general/general_content_000208.jsp
23 http://www.ema.europa.eu/ema/index.jsp?curl=pages/about_us/general/general_content_000264.jsp
24 https://www.cbd.int/abs/about/default.shtml
25 https://www.cbd.int/abs/bfmechanism.shtml
26 http://www.wipo.int/tk/en/igc/
27 http://www.wipo.int/export/sites/www/tk/en/resources/pdf/tk_brief2.pdf
28 http://plantauthority.gov.in/

Evidence-Based Medicine, Safety, and Efficacy

Clinical Trial Management in Botanical Drug Products

Ava Milani and Yashwant Pathak

Contents

12.1 Introduction

Advancing pharmaceuticals and alternative medicine is an area in science that is inordinately important and has recently received significant attention. Plant-derived products were the predominant source of medicine all over the world for centuries. As time passed, science and technology progressed and created what we are now familiar with—synthetic drugs. Standardization of these synthetic drugs helped cure and fight many diseases. Despite the major strides and majority of funding given to the development of synthetic drug production, we have recently seen a trend in the development of natural/botanical drug products. We have seen the trend of increasing use of herbal medicine go hand in hand with the passage of the Dietary Supplement and Health Education Act (DSHEA) [1]. The sale of herbal products has increased from 3.3 to 6.5 billion between 1990 and 1996 [2] and poised to be over 100 billion in 2020.

Botanical drug products, by definition given by the Food and Drug Administration (FDA), are products that include "plant materials, algae, macroscopic fungi, and combinations thereof [3]." Botanical drug products are prevalent in the market now. The botanical drug market itself had a value of $599 million in 2017 [2]. However, only two botanical drug products have been processed as prescription, FDA-approved drugs. While there are many botanical products currently under review by the FDA, we do not see as many botanical products as we would hope today, but in future the scenario will be changing.

This chapter specifically focuses on botanical drug products and their strengths and weaknesses. The chapter follows the FDA's recommendations on how to gain approval when submitting a product. The chapter will also discuss the relatively new term brought to the scientific world—evidence-based medicine—and then will continue to explore clinical trial management in botanical drug products.

12.2 Botanical drug products overview

The advancement of pharmaceuticals and alternative medicine is an area that has received much attention. Herbal medicine has been heavily used in many countries since the beginning of their history. Fossil records suggest that humans have used plants as medicine as far back as 60,000 years ago, in the Middle Paleolithic age [4]. It is said that the primary goals of using plants as therapeutic sources are as follows:

1. To isolate bioactive compounds for direct use as drugs
2. To produce bioactive compounds of novel or known structures as lead compounds for semi synthesis to produce patentable entities of higher activity and/or lower toxicity
3. To use agents as pharmacological tools
4. To use the whole plant or part of it as a herbal remedy [4]

Botanical drug products, also known as BDPs, are also often known as Traditional Chinese herbal Medicine (in China), TCM [5]. Botanicals are products that include "plant materials, algae, macroscopic fungi, and combinations thereof" [6]. In India it is part of the traditional Ayurvedic medicine system. In the ancient traditions and cultures (indigenous cultures), herbal medicine was the main resource for providing healthcare and disease treatment in almost all the countries. The World Health Organization (WHO) describes herbal medicine as drugs containing active ingredients plant parts or plant materials in the crude or processed state plus certain excipients [7]. One major difference between botanical products and standard western synthetic drugs is that botanicals often contain more than one major active ingredient [3]. This poses an issue for botanical products in many ways. Botanicals also do not produce immediate pharmacological results; therefore, they are rarely used in medical emergencies. Regardless of their lack of quick responses, botanicals are overall very suitable for chronic treatment.

There are several ways that a botanical product may be classified, including as a food, drug, medical device, or cosmetic. These four classifications were established under the Federal Food, Drug, and Cosmetic Act [5]. The United States National Institute of Health (NIH) is one of the many government run research organization focusing on the development of botanical drug products in hopes of producing more BDPs that can help cure life threatening diseases. As of now, the US Food and Drug Administration (FDA) has approved only two botanicals as prescription drugs. The first being Veregen (sinecatechins), which is a topical drug used to treat genital and perianal warts [8]. The second drug, which was approved in the end of 2012, 6 years after approval of the first BDP prescription, is Mytesi® (Fulyzaq®). Fulyzaq (crofelemer) is used to treat HIV/AIDS-related diarrhea, and it is made from the red sap of the *Croton lechleri* plant in South America [9].

12.3 The role of evidence-based medicine in botanical drug products

12.3.1 What is evidence-based medicine

In the 1990s a new term was to the scientific world—evidence-based medicine. Evidence-based medicine (EBM), is defined as "the conscientious, explicit, and judicious use of current best evidence in making decisions about the care of individual patients [10]." It entails a lifelong learning process that results in patients receiving the absolute best standard of care possible. One of the main reasons for the use of EBM is to use only the most current and relevant information [11]. Typical for any professional, especially those involved in health sciences, like physicians, the luxury of time is not always an option. A practicing physician is expected to read 19 scientific articles each day [10]. Most practitioners do not have this much time for additional reading, yet staying updated with recently published articles in medical journals is inordinately important for patient health care. The medical field changes daily with new technology and findings and therefore, keeping fully updated seems very difficult. EBM solves this issue, allowing physicians to read selectively and efficiently, making the best use of their precious time [12]. This term quickly rose in popularity, evident in many ways, such as the increasing number of scientific articles including the term.

12.3.2 Difference between evidence-based medicine and traditional medicine

It is vital to recognize the difference between evidence-based medicine and traditional medicine before exploring the role of evidence-based medicine on botanical drug products. Evidence-based medicine and traditional medicine both require a great deal of evidence. Evidence-based medicine requires a more in-depth proof of evidence than traditional medicine [13]. Evidence-based medicine requires more than just the opinion of a medical specialist or of any health professional, like what traditional medicine entails. It requires the medical specialist to integrate their experiences and medical knowledge with the most current and reliable evidence available [10]. To get a clear understanding of what evidence-based medicine really is, consider the following scenario, which includes a controversial procedure. The application of cricoid pressure (CP) is used to decrease aspiration when done during rapid sequence intubation (RSI), which is the most common method of intubation used today. CP is used during RSI when completing endotracheal intubation [14]. CP is very prevalent in anesthetic practice; however, it does not meet the standards for evidence-based medicine. CP does not meet the standards for appropriate randomized controlled trials, RCT, and, therefore, it is lacking proper evidence [15,16]. If the practitioner wanted to proceed with CP using evidence-based medicine, the only viable evidence would be that obtained from case series and case reports, the only evidence the physician would have on hand. For evidence-based medicine, that would be at the lowest strength for evidence, which is why the procedure of applying cricoid pressure is still controversial, even though it is widely used.

12.3.3 The role of evidence-based medicine in botanical drug products

With the increase in research surrounding botanical drug products, there has consequently been an increase in the amount of evidence-based research surrounding the topic as well. Research in botanicals is still quite controversial, mainly because of the concern by a number of scientists that clinical trials conducted of botanicals are of subpar quality. While that is one of the main concerns by scientists regarding botanicals, there are still many more concerns. The concern regarding the review process for botanicals is of high concern. Typically, when comparing typical Western medicine (synthetic products) to botanical drug products, the medical perception of Western medicine is more evidence based and botanicals are more experience base. The new role of evidence-based medicine has to be implicated upon the development of botanical drug products in order to be able to manufacture and produce a greater number of prescription botanical drug products in the United States.

12.3.4 Regulation of botanical drug products

In 2004 the FDA produced a Guidance for Industry on *Botanical Drug Products*. However, with the passage of time, there is a new understanding of Botanical Drug Products, and scientist and researchers have gained a greater experience, which has led to the FDA modifying the 2004 Guidance for Industry into a new and updated guide in the end of 2016. The new model includes detailed information on how to properly submit botanicals as marketed drugs [17]. One can submit a botanical drug product as a new drug application, NDAs, an investigational new drug application, INDs, as well as an over-the-counter (OTC) drug monograph system [17]. Due to the unique way of botanical drugs, there are differences in the way that synthetic, semi-synthetic, or chemically modified drugs are regulated and approved for marketing. The following sections will outline the current regulatory process for approval.

12.3.5 Submitting a botanical for FDA approval
12.3.5.1 Over-the-counter botanical drugs

A botanical product can fall under several classifications. For example, if the intended use of the botanical product is to treat disease, it would fall, by definition, under the classification of a drug. If the botanical product is intended to be produced as a dietary supplement, it would fall under the classification of a food [18]. If and when a botanical has an intended use that does not fall under the classification of a prescription drug (based on the US Food, Drug and Cosmetic Act) then the botanical is submitted as an over-the-counter drug 19]. Submitting a botanical as an OTC drug monograph requires the citizen to follow the Code of Federal Regulations (CFR) or a Time and Extent Application (TEA) [17] (pg. 3). Following those two codes, the botanical drug must also be recognized as an official United States Pharmacopeia and National Formulary (USP-NF) drug monograph [17] (pg. 3). If the botanical is not recognized by USP-NF then one would need to submit a "proposed standard for inclusion in an article to be recognized in an official USP-NF drug monograph" [17] (pg. 3). This very meticulous process makes this approach quite tedious. Therefore, it is suggested that one interested in submitting a drug as an OTC botanical drug to contact the Division of Nonprescription Drug Products in CDER's Office of New Drugs/ Office of Drug Evaluation IV for further information regarding the OTC drug monograph approach to marketing a botanical drug [17] (pg. 4). When submitting an application using a botanical product it is also suggested by the FDA to consult the FDA's Guidance for Industry of Botanical Drug Products [20].

12.3.5.2 Submitting an investigational new drug application (INDA) botanical drug

Submitting a botanical as an IND requires a great deal of evidence in proving that the drug is safe and effective. When submitting an IND, the FDA requires varying amounts of information regarding the drug depending on several factors, such as the amount of interaction the drug has had with human use, the number of reliable clinical studies that have been conducted on the drug, the drug's risks, and at what phase in development the drug is currently at [17] (pg. 8). As seen in the flow chart provided by the Guidance for Industry provided by the FDA (Figure 12.1), it is encouraged that the sponsor, who is the citizen submitting the drug, to provide all the available information on the drug, especially because an IND is submitted to the FDA during early phases of the drug's development, so as much data as possible is ideal at this point. The FDA encourages the sponsor to provide a description of the product and all documentation of prior human experience, chemistry, manufacturing, and controls, nonclinical pharmacology/toxicology, clinical pharmacology, and clinical considerations.

Submitting and achieving FDA approval of a botanical drug as a prescription drug is quite a daunting task for many—full of rules and regulations that make work in that department grueling and unappealing to some. Because of this, there are botanical products with an intended use as a medical treatment that are processed as a food supplement. The reason being, to resist going through the process of FDA approval. Many US manufacturers would rather use loopholes than try and receive FDA approval. Currently, as mentioned earlier in this chapter, there are only two botanical drugs produced as prescription drugs.

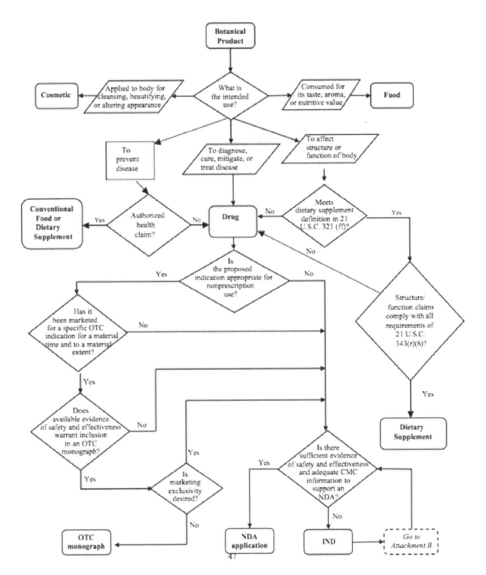

Figure 12.1 *Flowchart for botanical drug product development. (From Botanical Drug Development Guidance for Industry. [PDF File]. U.S. Department of Health and Human Services Food and Drug Administration, (CDER). Retrieved from http:// www.fda.gov/downloads/Drugs/Guidances/UCM458484.pdf, 2016.)*

12.3.5.3 New drug application for botanical drugs

When submitting a botanical product as a new drug application, there are several steps and measures that need to be taken to ensure the product is safe and reliable for human use. As you've seen on almost any label of ointments, drugs, etc., there is typically one active ingredient in Western medicine drugs.

In botanical drug products there are multiple active ingredients as well as inactive ingredients that all interact with each other in ways that are not always known [21]. This uncertainty results in a crucial contrast between botanical drug products and typical medical theory. The heterogeneity of the botanical drugs, as opposed to typical homogenous synthetic drugs, often results in batch-to-batch differences. These differences typically do not result in any meaningful differences in the therapeutic effects; however, it is imperative that the scientist thoroughly proves this in order to be approved as a NDA. The FDA suggests a totality-of-the-evidence path in proving the therapeutic effect is consistent [22]. Inconsistency of the drug may result in difference in effects in patients, or possibly resulting in adverse events for some. The totality-of-the-evidence approach includes several suggestions—botanical raw material control, quality control and manufacturing control, biological assay, and clinical data [17].

Medical theory consists of components such as mechanism and practice, diagnostic procedure techniques, criteria for evaluation of safety and efficacy in clinical trials, and treatment. For example, Western medicine provides a fixed dose that is typically universal among any patient who is taking the drug. The difference is usually due only to age and gender. Botanical drug products are given at a more flexible dose that varies between patients, based on medical history, age, and several other inclusive factors. All of these factors combined prove NDAs for botanical drug products to be quite difficult.

12.3.6 Clinical trial management in botanical drug products

12.3.6.1 Veregen (Kunecatechins) clinical trials

The first botanical drug product to be approved by the FDA is Veregen. The intended use of Veregen is to treat warts surrounding the genitals and anus [23]. The warts that Veregen targets are those that are caused by the human papillomavirus (HPV) [24].

The drug is made up of kunecatechins, which is a purified fraction of the water extract of green tea leaves from Camellia sinensis (L.) O Kuntze, which is a mixture of catechins and other green tea components [8]. The majority of the drug (by weight) is made up of catechins. Catechins make up 85%–95% of the drug's weight, and the rest of the drug is made up from gallic acid, caffeine, theobromine, and other unidentified constituents derived from the green tea leaves [8] (pg. 4). As expected, to ensure Veregen is a safe and effective drug, numerous clinical and nonclinical tests were conducted on the drug. According to the NDA, submitted for Veregan to the FDA, there were two phase III, randomized, double-blind, vehicle-controlled studies [8] (pg. 6). The studies were conducted on immunocompetent patients with external genital and perianal warts. The patients were required to be over the age of 18 and able to apply the Veregen ointment three times a day for up to 16 weeks, or until the warts were cleared. The sponsor provided the data, shown in Table 12.1 [8]. For the two studies, the median baseline wart area was 51 mm^2, and the median baseline number of warts was 6 [8].

Table 12.1 Efficacy Based on Region and Gender

Complete Clearance		Complete Clearance	
All Countries (includes the United States)		**Males**	
Veregen 15% (N = 21)	213 (53.6%)	Veregen 15% (N = 205)	97 (47.3%)
Vehicle (N = 207)	73 (35.3%)	Vehicle (N = 118)	34 (28.8%)
United States		**Females**	
Veregen 15% (N = 21)	5 (23.8%)	Veregen 15% (N = 192)	116 (60.4%)
Vehicle (N = 9)	0 (0.0%)	Vehicle (N = 89)	39 (43.8%)

Source: Frestedt, J. L., *Austin J. Nutr. Food Sci.* 5, 1–8, 2017.

12.3.6.2 Adverse events during Veregen clinical trials

The clinical trials intended in providing safety and efficacy of Veregen were successful, as shown by the positive results above. However, there were some adverse effects, which is important to note when studying botanical drug products [26]. Two subjects reported serious adverse effects. The patients reporting the adverse effects dealt with pain and inflammation where the Veregen was applied [27]. While those two patients dealt with serious adverse effects, 5% of the subjects also dealt with adverse effects, serious enough for it to cause the subjects to discontinue receiving the medication or to reduce the amount of dosage of Veregen. Severe reactions were seen in 37% of women and in 24% of men taking the Veregen. Common to clinical trials of any drug, there were many different reactions seen in the subjects, some related to the drug, and some possibly unrelated [8].

12.4 Fulyzaq (Crofelemer) clinical trials

In December of 2012, the FDA approved the second botanical drug product, Fulyzaq (crofelemer). Fulyzaq is manufactured by a US-based pharmaceutical company called Salix Pharmaceuticals. This drug has proved to be quite successful in treating non-infectious diarrhea in HIV/AIDs patients who are on antiretroviral therapy (ART), and it is currently the only drug that can treat non-infectious diarrhea in HIV/AIDs patients [28]. While there have been major strides in HIV control, the major adverse effect of ART is diarrhea, which can cause weight loss, cramps, dehydration, and a major decrease in the quality of one's life. The quality of life for some diminishes so drastically when on antiretroviral therapy, that it urged many people to stop their HIV/AIDs treatment of ART early or switch to other treatment plans. Fulyzaq was able to solve this problem.

It was found that chloride secretion was the primary cause of diarrhea in HIV patients. Fulyzaq inhibited chloride secretion in the gastrointestinal lumen, reducing diarrhea and also reducing water loss, which is another main cause of diarrhea [22] The clinical trial that the FDA based their approval on was known as ADVENT. This study was a randomized and double-blind, 48-week Phase III clinical trial, conducted between October 2007 and December 2011 [28]. It included over 690 patients, with about half of them HIV patients undergoing antiretroviral therapy with a diarrhea side effect. In the end, the patients who were administered the Fulyzaq as opposed to the placebo saw promising results; a reduction in the number of watery bowel movements and stool consistency score during a day. While this was a major study in the success of Fulyzaq, it was not the only one [29]. Fulyzaq underwent numerous clinical trials. Some of which will be discussed [30].

12.5 Fulyzaq related clinical trials

Crofelemer, the main ingredient in Fulyzaq, has been tested in many clinical trials, proving its effectiveness. In understanding clinical trial management of botanical drug products, it helps to review examples of the types of clinical trials these botanicals are conducted in before it is produced and manufactured. Sponsored by Salix Pharmaceuticals, the company that currently manufactures Fulyzaq conducted a study using the drug crofelemer to test the effectiveness of crofelemer in treating women with diarrhea-predominant irritable bowel syndrome [31]. This study required participants to meet several criteria, much like any study. Most importantly, the females needed to be diagnosed with diarrhea predominant irritable bowel syndrome. There are many other criteria necessary to be involved in the trial. The official title of the study is a randomized, double-blind, placebo-controlled study [31]. This was a four year-long study. Another study, also conducted by Salix Pharmaceuticals, had an outcome for serious adverse effects and an incidence of treatment-emergent adverse events. This study was a phase III, multicenter, open-label evaluation of the safety and tolerability of crofelemer in HIV-positive subjects with diarrhea [32]. The participants were given 125 mg of crofelemer orally, two times a day. During clinical trials, scientists were faced with adverse effects. The most common adverse effects included infections and GI disorders. However, in the ADVENT study, these adverse effects were similar during the time the patients were receiving crofelemer as well as during the placebo period.

12.6 How to optimize clinical trial budgeting in botanical drug products

Depending on many factors, like the complexity of what is being studied, clinical trials can end up being quite complex and intense weeks of study. This makes clinical trial budgeting inordinately important to be able to manage the trial as well as meet all criteria and guidelines, especially when conducting clinical trials that include botanical drug products [33]. The National Center for Biotechnology Information (NCBI) developed a standard model that allows those conducting clinical trials to score their proposed trial a "value/score" in hopes of receiving appropriate effort from their participating sites [34]. The model includes 10 different parameters, each ranked according on its level of complexity: routine, moderate, or high. The parameters are as follows:

1. Number of study arms/study groups
2. Informed consent process
3. Enrollment feasibility/study population
4. Registration of study participants and randomization process
5. Nature of investigational product and complexity of administration
6. Length of investigational treatment phase
7. Study team's/study staffs
8. Data collection complexity
9. Follow-up requirements
10. (A and B) Ancillary studies [34]

The researcher should then go through these parameters and assign points to each. The points are assigned as follows: 0 points for routine, 1 point for moderate, and 2 points for high. (See reference 17 for more information.)

The FDA requires a great deal of information regarding any drug when one is seeking approval, but especially in-depth information surrounding botanical drug products. However, there are some differences in clinical study requirements for BDPs because there is typically more than one chemical constituent in botanical products [17]. Generally, the FDA states that the requirements for in vivo bioavailability data in an NDA is applicable to BDP. This includes "information on the active constituents, the complexity of the drug substance, and the availability of sensitive analytic methods" [17] (pg. 28). However, the bioavailability data in an NDA can be waived, due to the inability or difficulty to perform standard in vivo bioavailability.

12.7 Safety and efficacy of botanical drug products

For as long as humans have existed, herbal and natural products have been used for many reasons, like treating illness. While not every all-natural product is safe, botanical drug products have proven to be quite successful in many cases [19]. In the following sections, we will discuss the safety and efficacy of these products as well as the threats and adverse effects that stand with botanical drug products.

12.8 The issue of anecdotal evidence

There is a growing need of evidence-based medicine in botanical drug products. One major downfall of the amount of evidence we currently have on the safety and evidence of botanical drug products is that majority of evidence is anecdotal. Anecdotal evidence is regulated in a modern clinical setting, with appropriate research, tools, and technology; however, this does not mean that anecdotal evidence is not helpful. But before exploring anecdotal evidence of botanical drug products, understand the reasons for anecdotal evidence providing some issues for development.

Historically, botanical medicines are documented from folk traditions, meaning there is not a great deal of written information regarding them. While there is not as much written evidence wanted about botanical products, majority of developing countries rely on plant products as their primary form of health care [35]. Botanical medicines are quite difficult to standardize, mostly because they are parts of raw botanical material or extracts of either a whole plant or just parts of it [36]. Due to their nature, this makes the time of harvest, the method of extraction (there are many different methods of extraction), and the plant's growing conditions difficult to standardize, which is something Western medicine is used to.

Another major issue of dealing with anecdotal evidence arises from the fact that botanical products have existed for so long, and not been regulated. An example helps understand this situation. For hundreds of years, people native to Hawaii have been drinking Kava, a crop native to the land. Kava was used in social and religious settings, and the native Hawaiians consuming it were perfectly used to it and healthy after consumption. However, there are many cases of people in the United States and Europe dealing with adverse effects from Kava, such as hepatotoxicity. This shows that when humans co-evolve with these medicinal plants, their data surrounding that plant may not be accurate for people native to other land: healthy results in one populations, but toxic effects in another.

The issue of anecdotal evidence does not rule out other types of evidence when dealing with botanical drug products; it is just reason for the FDA providing many rules and regulations as well as a detailed guideline for botanical drug approval [37].

12.9 Safety of botanical drug products

Crop-to-crop variation of plants makes manufacturing botanical products difficult. These crop-to-crop changes are due to natural changes in growing conditions, like soil conditions. Another issue is once the herbs being used in the botanical product are processed, other natural elements, like air, ostrure, and light, can all affect the potency and value of the botanical products [17]. Because of all of these safety issues, the FDA has provided recommendations on how to deal with the botanical raw materials used in the product to ensure safety and efficacy. The FDA would like the sponsor to provide information regarding "identification of the plant species, plant parts, alga, or macroscopic fungi used, a certificate of authenticity, whether the plant species is determined to be endangered or threatened under the Endangered Species Act or the Convention on International Trade in Endangered Species of WIld Fauna and Flora, entitled to protection under some other federal law or international treaty to which the United States is a party, and/or in a critical habitat that has been determined to be endangered or threatened [17] (pg. 8). The FDA also asks for information regarding each grower and/or supplier such as name and address, description and characterization of the plant species (as well as varieties and cultivars and botanical identification, harvest location, growth condition, stage of plant growth at harvest, and harvest time/season, and post-harvesting processing, control of foreign matter, preservation procedures, tests for elemental impurities, tests for residual pesticides, and tests for adventitious toxins [17]."

12.10 Safety-thorough regulation necessary

When a botanical drug is produced or made available to the US market, it will be considered as a "new" drug. A new drug, as defined by the FDA, is "...not generally recognized as safe and effective under the conditions prescribed, recommended or suggested in the labeling" (Botanical Drug Development Guidance for Industry (2016) Food and Drug Administration). This ensures that the botanical, which may already have been used for years, undergoes additional testing to ensure safety. The FDA often will require more clinical and nonclinical information to support a particular indication, novel route, or new population [38]. As explained earlier in the chapter, the sponsor always needs to first submit an Investigational New Drug Application (IND) and go through the process of then submitting a New Drug Application (NDA) and wait for approval before producing and manufacturing the botanical drug product to the US market.

12.11 Efficacy of botanical drug products

Proving efficacy in the US market for botanical drug products is dependent on the regulatory route that the sponsor has taken. An article named "Botanical Regulation: A Comparison of the United States and Canada" has provided four principles to help define US regulation. They are:

Route of administration
Form or formulation

Safety
Intended use [39].

Due to the unique nature of botanicals, there are several ways that they can be classified. Different classifications require different routes of proving efficacy. Many botanicals are produced and sold as conventional foods, dietary supplements, cosmetics, or as medical devices [40]. For these classifications it is said that the idea of proving efficacy is not applicable. These products must be in alignment with precedent devices under Section 510 k of the FDC act.

Proving efficacy for botanical products that will be produced as drugs requires virtually the same amount of information and testing as any other FDA approved drug. Described earlier, there are many differences in testing and approval from the FDA when dealing with botanical products, however overall the process is quite similar to any other drug. The FDA will only conclude a drug as safe and effective for its intended use when it is defined as having "evidence consisting of adequate and well-controlled investigations, including clinical investigations, by experts qualified by scientific training and experience to evaluate the effectiveness of the drugs involved [27]."

The FDA has a well-developed system for approving botanical products as botanical prescription drugs; however, botanical products that act as conventional foods, dietary supplements, cosmetics, or medical devices are not advertised and promoted like prescription drugs are by the FDA [40]. Botanicals not classified as prescription drugs require the US Federal Trade Commission to advertise and promote.

12.12 Issues in developing botanical drug products

Standardization in botanical drug products is often quite hard to produce, due to the unique nature of the drugs. As explained earlier, botanical drug products typically contain multiple active constituents. These constituents are generally unidentifiable, isolated, not chemically defined, along with that, they also lack a well-characterized full spectrum of biological activity [41] (pg. 667). These factors, combined, make standardization difficult; however, without knowing what the active constituents are and their biological activity, one is not able to ensure the product safe, making standardization a difficult but important factor in botanical drug product safety [42].

Another major issue in developing botanical drug products is another standardization issue. While the botanical products can be derived from the same species, they may have significant differences that alter their strength/potency [41]. The botanicals may contain impurities that are a result of the difference in geographic location, the time of harvest, and the part of the plant that is used [41] (pg. 667). Standardization of chemicals in drugs is inordinately important, but the worth of the botanical product is hindered when the part of the plant, or material, that is used is not well-characterized. Harvesting, processing, and formulation variations can also drastically alter the quality and consistency of the product because the differences in processing may invite new contaminants into the final product.

12.13 Safety concerns of botanical drugs

For a botanical drug to be deemed safe by the FDA, nonclinical and clinical safety evaluations must be taken place. It is vital to follow the FDAs suggestions from the "Botanical Drug Development Guidance for Industry" when submitting INDs and NDAs for the botanical product. The Center for Drug Evaluation and Research (CDER) current development plans of botanical drugs provides specific recommendations that one would find helpful in developing a botanical product. When there are serious adverse effects of the botanical product, safety concerns are brought up. The FDA has total control over approval of the product and may put the development plans on hold when an IND is submitted without appropriate information given [41] (pg. 672).

The level of safety evaluation testing for the botanical product differs depending on the phase of development the product is in. Less information is needed for early phase studies, and much more in-depth information is needed for later phase studies. When a botanical product is initially submitted for approval, the FDA will treat and determine approval of the drug as they would any other New Chemical Entity (NCE) that is under development. Early phases of BDP development require nonclinical pharmacology and toxicology studies to prove safety. Previous human experience alone does not fulfill the safety needs, nor does it fulfill enough data needed for chronic therapy. The FDA considers safety data from other countries while determining the level of importance of other nonclinical safety evaluations.

12.14 Nonclinical pharmacology/toxicology information of botanical drug products

Submitting applications for botanical drug products, whether it be an IND or a NDA, mainly depends on the amount of information available on the product. The FDA may require less information, especially during the early phases of production, when there is a great amount of previous human use with the botanical. The previous human use can prove safety of the product. The FDA values human experience with the botanical most when the proposed botanical is submitted using the same dosage and duration as humans previously used it. The FDA evaluates the amount of nonclinical information necessary depending on the situation at hand. There are several different routes the FDA may take. The following examples are described in the Botanical Drug Development Guidance for Industry.

Many botanicals produced in the US are produced as dietary supplements, due to difficulty gaining FDA approval as a prescription drug, or an over-the-counter drug. If the botanical is currently produced as a dietary supplement, the FDA may decide to not request further nonclinical pharmacological/ toxicological testing. The dietary supplement must be legal and lawfully marketed in the United States (1, pg. 13). The FDA's decision to request less information is contingent on the dietary supplement having the same intended use and dosage as the proposed use of the future botanical drug product [17].

Another exception where the FDA will require less nonclinical pharmacological/ toxicological testing is for a botanical drug with a great deal of past human experience. The botanical drug does not have to be currently manufactured in the US; however, if the drug is administered using the route, and prepared, processed, and used according to methodologies with prior human use, the FDA will find that information enough (1).

The FDA does not make any exceptions in the required nonclinical pharmacological/toxicological testing if the botanical drug dosage in the proposed clinical studies is greater than its dosage in prior human use [17]. For example, if the proposed study explained the botanical as having a higher dose or longer duration, the FDA will require sufficient testing to prove safety and efficacy.

12.15 Toxicity studies of botanical drugs

Generally, two types of toxicity studies are important when developing safe and effective botanical drug products. These two types are acute toxicology studies, and repeat-dose toxicity studies.

Acute toxicology studies are not always relevant when developing botanical drug products. Because many botanicals that are being developed have been used by humans for many year prior, the amount of previous human experiences acts as a safety study making the acute toxicology study unnecessary. This makes acute toxicology studies not important when developing a botanical drug product as an IND.

The other toxicity study is repeat-dose toxicity studies. When an NCE is submitted to the FDA, a repeat-dose toxicity study is also required. The study entails evaluation of two mammalian species, ideally a rodent and a non-rodent species. Then, a high dose of the product is used to produce a toxic effect in the animal. The scientist should aim to test the toxicity using the amount of product that would be proposed for clinical use. The study should last at least as long as the proposed clinical trial, which is usually at least two weeks, this rule is in accordance with the ICH Guidance M3 (R2) Nonclinical Safety Studies for the Conduct of Human Clinical Trials for Pharmaceuticals. It is also important to note that the study should not last longer than nine months in the non-rodent species and should not last longer than six months in the rodent species [41].

While there are two major toxicity studies that should be conducted during development phases of botanical products, if studies show that certain organs or organ systems are being targeted during toxicity studies, then special toxicology studies must be conducted. This include in vitro and in vivo special toxicity studies [43]. These are necessary to explain the reason for the toxicity in those organs and to ensure safety [44].

12.16 Genetic toxicity testing of botanical drug products

Typically, genetic toxicity testing is required, by law, for all chemicals and drugs. However, it's not always required on a compound to compound basis, for a variety of reasons [45]. For developing of botanical drug products, most of which have been used by humans for a number of years, the FDA does waiver this law for some botanicals [46]. While genetic toxicity testing is not always required during the IND phase of BDPs, to fulfill final NDA requirements, standard genetic toxicity testing should be completed. A Comprehensive Guide to Toxicology in Preclinical Drug Development provides three tests that should be completed for genetic toxicity testing. They are as follows: "a test for gene mutation in bacteria, an in vitro test with cytogenetic evaluation of chromosomal damage with mammalian cells or an in vitro mouse lymphoma TK assay, and an in vivo test for chromosomal damage using rodent hematopoietic cells" [41] (pg. 672). If one of the following test shows positive findings, the sponsor should find alternative ways to prove genetic toxicity as not an issue in safety of the botanical drug product's final phase.

In botanical drug product development there has been a trend in the awareness of the importance of genotoxicity testing as well as an increase in the priority of genotoxicity testing by sponsors when developing botanical drug products. These results are concluded after a survey conducted by the FDA [46]. The survey studied botanicals during the IND phase of development from 2001 to 2008. Genotoxicity testing is vital to development and is found to be more cost effective compared to other tests during the development phases of BDP production. The FDA urges sponsors to retrieve information regarding genetic testing as early as possible and aim to have the information before the introduction of humans into the clinical trials [41] (pg. 672).

12.17 Developmental and reproductive toxicity studies

When fulfilling the requirements of an NDA, clinical trials testing development and reproductive toxicity are also quite important [25]. These studies, referred to as DART studies, are essential to clinical trials to provide safety evidence for potential issues, like problems with fertility, problems with embryonic development, and teratology in both the rodent and non-rodent species being tested. According to "A Comprehensive Guide to Toxicology in Nonclinical Drug

Development [47]," "Dart studies are designed to identify the effects of drugs on mammalian reproduction and include exposure of mature adults, as well as all stages of development from conception to sexual maturity." There is not a standard of testing for DART studies because the requirements will most likely differ depending on the patient being tested with the botanical drug product [41].

12.18 Carcinogenicity studies of botanical drug products

Along with all the other necessary testing that a sponsor should conduct when developing a botanical drug product, carcinogenicity testing is also one that is extremely important. This is needed only when the intended duration of treatment with the botanical product is longer than three to six months [48]. If the botanical product is only intended to be used for short amount of time, such as less than two weeks like in antibiotic therapies, carcinogenicity testing is irrelevant [48]. However, when it is needed, the FDA suggests that at least two carcinogenic tests be conducted [41].

12.19 Critical control points of botanical drug products

We have seen many strides in the development of botanical drug products. Unfortunately, due to tough regulations and weak incentives for production, their development sees serious limitations. Between 1999 and 2012, over 500 IND applications were submitted to the FDA, and only two prescription botanical drug products were approved [49]. There are several reasons for this lack of progress in BDP development, some of which will be explained in the following sections.

12.20 Issues with gaining approval of botanical drug products

An extensive and detailed guidance for developing drugs seems fair. In the case of botanical drug products, it often creates an inability for sponsors to get their safe and effective drugs approved. In the United States, the FDA is very well respected, and its approval of food, and especially drugs, is vital to US consumers [36]. Without FDA approval, US consumers are hesitant to embrace botanical medicine, creating a greater case against a less intense environment for the BDP drug approval process [1]. In December of 2016, the FDA revised the 2004 version of the Botanical Drug Development Guidance for Industry. The revised version provides specific recommendations and details regarding how to gain FDA approval. The FDA's revised guidance on BDPs requires less information during the IND phase, but the same information is still required for the NDA phase [21]. The FDA was able to ask for less information during the IND phase because botanical products have a great prevalence in being used in treatment of human diseases throughout history [2]. Pharmaceutical companies are still confused by the guidance and are still having trouble with submitting their drug applications [2]. The major issue sponsors have with the guidance on botanical drugs is moving from an IND into an NDA. Sponsors find submitting an IND relatively easy, mostly because it does not require too much in depth information surrounding the drug (see Submitting an Investigational New Drug Application Botanical Drug, Section 12.3.5.3). The NDA is where the FDA will evaluate safety, efficacy, proper labeling, and proper manufacturing methods; therefore it is evaluated at a much higher intensity, making approval quite difficult [2].

12.21 Conclusion

Botanical drug products can help treat and cure many life-threatening diseases. Pharmaceutical companies all over the United States are becoming more and more aware of their positive effects. Several pharmaceutical companies are currently working toward gaining FDA approval; however, as explained previously, FDA approval can be quite tough. As of now, the FDA has approved two prescription botanical drug products. These two drugs, Veregen and Fulyzaq, gained FDA approval and proved to be successful. It is imperative that more pharmaceutical companies work toward gaining FDA approval of their botanical products in order for their use to be more trusted and accessible to the US consumer.

References

1. Li, W., Botanical drugs: The next New New thing? (2002 Third Year Paper).
2. Ahn, K. (2017). The worldwide trend of using botanical drugs and strategies for developing global drugs. *BMB Reports*, 50(3): 111–116. doi:10.5483/BMBRep.2017.50.3.221.
3. Lee, S. Botanical drug development and quality standards. US Food and Drug Administration, October 6, 2015. Accessed on January 10, 2018. Retrieved from http://pqri.org/wp-content/uploads/2015/10/01-PQRI-Lee-Botanicals-20151.pdf.
4. Katiyar, C., Gupta, A., Kanjilal, S., & Katiyar, S. (2012) Drug discovery from plant sources: An integrated approach. Ayu, 33(1), 10–19. doi:10.4103/0974-8520.100295.
5. Hoffman, F. A. The state of botanical drugs. Accessed on January 11, 2018. Retrieved from http://www.heterogeneity-llc.com/uploads/5/1/0/5/51059387/hoffman_fa_-botanical_drugs_in_us_nutraceuticals_world_06_web.pdf.
6. Miroddi, M., Mannucci, C., Mancari, F., Navarra, M., and Calapai, G. (2013) "Research and development for botanical products in medicinals and food supplements market," *Evidence-Based Complementary and Alternative Medicine*, 2013, Article ID 649720, 6. doi:10.1155/2013/649720.
7. Calixto, J.B. (1999) Efficacy, safety, quality control, marketing and regulatory guidelines for herbal medicines (phytotherapeutic agents). Retrieved from http://www.scielo.br/pdf/bjmbr/v33n2/3704c.pdf.

8. Veregen™ [PDF File]. pp. 4–17. Retrieved from https://www.accessdata.fda.gov/drug-satfda_docs/label/2006/021902lbl.pdf.
9. McKenna, D. Is there good scientific evidence? Accessed on January 15, 2018. Retrieved from https://www.takingcharge.csh.umn.edu/explore-healing-practices/ botanical-medicine/-there-good-scientific-evidence.
10. Palmer, E. (2012). FDA approves 2nd botanical, Fulyzaq from Salix. Accessed on January 14, 2018. Retrieved from https://www. fiercepharma.com/regulatory/ fda-approves-2nd-botanical-8-years-fulyzaq-from-salix.
11. Haughom, J. 5 Reasons the practice of evidence-Based medicine is a hot topic. Retrieved from https://www.healthcatalyst.com/5-reasons-practice-evidence-based-medicine-is-hot-topic.
12. Masic, I., Miokovic, M., & Muhamedagic, B. (2008). Evidence based medicine—New approaches and challenges. *Acta Informatica Medica*, 16(4): 219–225. doi:10.5455/ aim.2008.16.219-225.
13. Le, Hong Phuong Nhi. (2016). Eminence- Based medicine vs. evidence-Based medicine. Accessed on January 15, 2018. Retrieved from http://www.students4bestevidence.net/ eminence-based-medicine-evidence-based-medicine/.
14. Lal, R. (2015). Botanical drug review. *FDA/CDER SBIA Chronicles*, 1–2.
15. Stewart, J. C., Bhananker, S., & Ramaiah, R. (2014). Rapid-sequence intubation and cricoid pressure. *International Journal of Critical Illness and Injury Science*, 4(1): 42–49. doi:10.4103/2229-5151.128012.
16. Katz, R. (2004). FDA: Evidentiary standards for drug development and approval. *NeuroRx*, 1(3): 307–316.
17. (2016). Botanical Drug Development Guidance for Industry [PDF File]. U.S. Department of Health and Human Services Food and Drug Administration, Center for Drug Evaluation and Research (CDER). Retrieved from http://www.fda.gov/down-loads/Drugs/Guidances/UCM458484.pdf.
18. U.S. Food and Drug Administration. What is a botanical drug? Retrieved from https:// www.fda.gov/AboutFDA/CentersOffices/OfficeofMedicalProductsandTobacco/ CDER/ucm090983.htm.
19. Schmidt, B., Ribnicky, D. M., Poulev, A., Logendra, S., Cefalu, W. T. & Raskin, I. (2008). A natural history of botanical therapeutics. *Metabolism: Clinical and Experimental*, 57(7 Suppl 1): S3–S9.
20. Schiff, R., (2009). Botanical products: Drugs and dietary supplements [PDF File]. Accessed on January 5, 2018. Retrieved from www.raps.org/WorkArea/DownloadAsset. aspx?id=3862.
21. (2012). Botanical drug market hampered by regulations. Accessed on January 5, 2018. Retrieved from https://www.naturalproductsinsider.com/news/2012/12/botanical-drug-market-hampered-by-regulations.aspx.
22. Larry, S. L. (2015). *Botanical Drug Development and Quality standards [PDF]*. U.S. Food and Drug Administration, Silver Spring, MD.
23. (2015–2018). Veregen ointment. Accessed on January 10, 2018. Retrieved from https:// www.webmd.com/drugs/2/drug-149848/veregen-topical/details.
24. Chow, S. C., Pong, A (2015) Scientific issues in botanical drug product development. *Annals of Biometrics & Biostatistics*, 2(1): 1012.
25. Frestedt, J. L. (2017). Similarities and difference between clinical trials for foods and drugs [PDF File]. *Austin Journal of Nutrition and Food Sciences*, 5(1): 1–8.
26. Bent, S. (2008). Herbal medicine in the United States: Review of efficacy, safety, and regulation: Grand rounds at University of California, San Francisco Medical Center. *Journal of General Internal Medicine*, 23(6): 854–859. doi:10.1007/ s11606-008-0632-y.
27. Fulyzaq (Crofelemer) for the treatment of HIV/AIDS-Associated diarrhoea. Accessed on January 8, 2018. Retrieved from http://www.drugdevelopment-technology.com/ projects/fulyzaq-crofelemer-treatment-hiv-aids-associated-diarrhoea/.

28. Framptom, J. E. (2015). Crofelemer: A review of its use in the management of non-infectious diarrhoea in adult patients with HIV/AIDS on antiretroviral therapy. Accessed on January 11, 2018. Retrieved from http://www.druglib.com/abstract/fr/frampton-je1_drugs_20130000.html.

29. (2015). A double-blind, randomized, placebo-controlled, multicenter study to assess the efficacy and safety of orally administered sp-303 for the treatment of diarrhea in acquired immunodeficiency syndrome (aids) patients. Accessed on January 9, 2018. Retrieved from http://www.druglib.com/trial/08/NCT00002408.html.

30. (2015). Diarrhea predominant irritable bowel syndrome in females. Accessed on January 11, 2018. Retrieved from http://www.druglib.com/trial/26/NCT00461526.html.

31. (2015). Safety and effectiveness of 3 doses of crofelemer compared to placebo in the treatment of HIV associated diarrhea. Accessed on January 12, 2018. Retrieved from http://www.druglib.com/trial/98/NCT00547898.html.

32. (2015). Safety and tolerability of Crofelemer for HIV-Associated diarrhea. Accessed on January 11, 2018. Retrieved from http://www.druglib.com/trial/90/NCT01374490.html.

33. Malikova, M. A. (2016). Optimization of protocol design: A path to efficient, lower cost clinical trial execution. *Future Science OA*, 2(1), FSO89. doi:10.4155/fso.15.89.

34. Adams, L. (2016). Botanical medicines in research [Whitepaper]. 2018, January 16, from http://go.quorumreview.com/rs/169-RVQ-430/images/08_Quorum_WP_BotanicalMedicines_080216.pdf.

35. Calixto, J. B., (2000). Efficacy, safety, quality control, marketing and regulatory guidelines for herbal medicines (phytotherapeutic agents). *Brazilian Journal of Medical and Biological Research*, 33(2): 179–189.

36. Zhang, A. L., Xue, C. C., & Fong, H. S. H. (2011) Integration of herbal medicine into evidence-based clinical practice. In Benzie IFF, Wachtel-Galor S (Eds.), *Herbal Medicine: Biomolecular and Clinical Aspects*. (2nd ed.) Boca Raton, FL: CRC Press/Taylor & Francis Group.

37. Developing a Strategy to Optimize Clinical Trial Supplies. (2014). [ebook] mediate, pp. 1–6. Available at: https://www.mdsol.com/sites/default/files/product-files/BAL_Developing-Strategy-Optimize_20141103_Medidata_White-Paper.pdf (Accessed January 16, 2018).

38. Fabricant, D. S., Farnsworth, N. R. (2001) The value of plants used in traditional medicine for drug discovery. *Environmental Health Perspective*, 109(Supplement 1): 69–75.

39. Tamayo, C., Hoffman, F. A., (2017) Botanical regulation: Comparison of the United States and Canada. *pharmacy Regulatory Affairs*, 6: 189. doi: 10.4172/2167-7689.1000189.

40. Dietary Supplements. U.S. Food & Drug Administration. June 6, 2018. Accessed on July 20, 2018. Retrieved from https://www.fda.gov/food/guidanceregulation/guidancedocumentsregulatoryinformation/dietarysupplements/ucm073200.htm.

41. Tamayo, C., & Hoffman, F. A. (2017). Botanical regulation: Comparison of the United States and Canada. *Pharmaceutical Regulatory Affairs: Open Access*, 6(1): 1–5.

42. George, P. (2011) Concerns regarding the safety and toxicity of medicinal plants—An overview. *Journal of Applied Pharmaceutical Science*, 01(06): 40–44.

43. Faqi, A. S., Hoberman, A., Lewis, D., & Stump, D. (2017) Developmental and reproductive toxicology. In Faqi, A. (Ed.), *A Comprehensive Guide to Toxicology in Nonclinical Drug Development*. London, UK: Haley, Mica. pp. 216–241.

44. Karimi, A., Majlesi, M. & Rafieian-Kopaei, M. (2015) Herbal versus synthetic drugs; Beliefs and facts. *Journal of Nephropharmacology*, 4(1): 27–30.

45. Faqi, A. S., & Yan, J. S. (2013). Nonclinical safety assessment of botanical products. In *A Comprehensive Guide to Toxicology in Preclinical Drug Development*. Amsterdam, the Netherlands: Academic Press. pp. 665–674.

46. Wu, K. M., Dou, J., Ghantous, H., Chen, S., Bigger, A., Birnkrant, D. Current regulatory perspectives on genotoxicity testing for botanical drug product development in the U.S.A. *Regulatory Toxicology and Pharmacology*, 56(1): 1–3. doi:10.1016/j.yrtph.2009.09.012.

47. (2018) Genetic toxicology. Retrieved from https://www.criver.com/products-services/safety-assessment/toxicology-services/genetic-toxicology?region=3601.

48. Miroddi, M., Mannucci, C., Mancari, F., Navarra, M., & Calapai, G. (2013). Research and development for botanical products in medicinals and food supplements market. *Evidence-Based Complementary and Alternative Medicine: ECAM, 2013*, 649720. doi:10.1155/2013/649720.

49. Hoffman, F. A., & Kishter, S. R. (2013). Botanical new drug applications—The final frontier. *American Botanical Council*, 10(6).

Index

Note: Page numbers in italic and bold refer to figures and tables respectively.

PPVFR (Protection of Plant Varieties and Farmers' Rights) 241
preclinical characterization: crofelemer (Fulyzaq) 174; ERr 731 extract (Estrovera) 179; Sinecatechins (Veregen) 170
preparative column chromatography 101–3, *102*
prescription query 128
proanthocyanidins *175*
profit probability 8
progress period 8
ProPass™ stent 146
Protection of Plant Varieties and Farmers' Rights (PPVFR) 241
PubChem 65
Purusha-Prakriti Siddhanta 80–1
Pushpangadan, P. 55

quantitative composition–activity relationship (QCAR) 127

rapid sequence intubation (RSI) 248
rare disease patient populations 187
Rasa Shastra/Bhaishajya Kalpana 84
R&D, *see* research and development (R&D)
Regenera Pharma 193
remote triggers 211
repeat-dose toxicity 260
research and development (R&D): cost 8–10; pharmaceutical 9–10
Retina Implant AG 147
reverse pharmacology (RP) 62, 124; botanical drug development *125*; and forward strategy *63*; ladder *64*; natural compounds database 63–71; role of 63–73; virtual screening 71–3
Rhyncholacis penicillata 48–9
RIKEN NPEdia 71
RSI (rapid sequence intubation) 248
ruthenium 141

salix pharmaceuticals 253
Samhita medicinal literature 83
SANCDB database 67
Sanfujiu treatment 108–9
sangre de drago 174
Sapta-Padartha Siddhanta 81
Sativex 74
Saudi Arabia 237
SBVS (structure-based virtual screening) 71
SCF (Supercritical fluid-based) manufacturing techniques 146
Science 41

scientific literature analysis 121
Second Sight Medical Products Inc 147
secretory diarrhea 176
selective estrogen receptor modulator (SERM) 178
sertraline 220
sexually transmitted infection (STI) 168
shape methods 73
Shat Kriya Kaala/Roga Marga 83
sheep thyroid 96
shell region 216
shot-gun/random approach 49–50
Siberian rhubarb 179
Sienna+® 147
SIFt (structural interaction fingerprint) method 72
silica nanoparticles 152
Sinecatechins (Veregen) **167**; chemistry 169–70; clinical pharmacology 170–1; preclinical characterization 170; standard indication 168–9; totality of evidence 172
solubility factors 93–4
SPC (supplementary protection certificate) 230
spectroscopic techniques 211
SpiceRx 70
Spirituality 41
Srotasa/Maheshwara Sutra 82
stealth liposomes 211
STI (sexually transmitted infection) 168
Story of My Botanical Studies (book) 46
stratum corneum 219
structural interaction fingerprint (SIFt) method 72
structure-based virtual screening (SBVS) 71
subcritical water extraction (SWE) 99–100
Super Natural II 65
supplementary protection certificate (SPC) 230
surface erosion 218
surgery 148–9
SWE (subcritical water extraction) 99–100

target query 128
TBGRI (Tropical Botanic Garden and Research Institute) 54–5
TCM (Traditional Chinese Medicine) 28–9, **30**, 118, 234
TCMGeneDIT database 68
TCMID (Traditional Chinese Medicine Integrated Database) 65, 128, 132
terpenoids 119
tetrahydrocannabinol (THC) 119